Faith and Reason

The Universality of God
and Fallacy of Atheism

Dr. Jim I. Jones

Copyright © October 2017 by Dr. Jim I. Jones

All rights reserved. No part of this book may be reproduced or transmitted in any form or by any means, electronic or mechanical, including photocopying, recording, or by any information storage or retrieval system without written permission from Dr. Jim I. Jones, except for the inclusion of quotation in review.

Judith G. Jones photographed the cover

Printed in the United States of America

ISBN: 1-970024-75-5

ISBN-13: 978-1-970024-75-3

Library of Congress Control Number: 2017914553

Publisher: Dr. Jim I. Jones

**Dedicated to:
Father Joe Hopkins,
Society of Jesus**

Father Joe and I spent many hours at local pubs where he labored to reduce the depth and breadth of my ignorance on the universality and authority of God.

- *The Stanford Encyclopedia of Philosophy* describes faith in a number of relationships to reason:
- Faith underlies reason: all human knowledge and reason is seen as dependent on faith: faith in our senses, reason, memories, and accounts of events from others.
- Faith addresses issues beyond the scope of reason: faith covers issues that science and rationality are incapable of addressing, but that are real.
- Faith and reason are essential together: The Catholic view is that faith without reason leads to superstition, while reason without faith leads to relativism.
- Faith based on warrant: some degree of evidence provides warrant for faith.
- Faith contradicts reason.

The Relationship Between Faith and Reason in an Encyclical Letter from Pope John Paul II

"…Complex systems of thought have thus been built, yielding results in the different fields of knowledge and fostering the development of culture and history... (Yet) reason, in its one-sided concern to investigate human subjectivity, seems to have forgotten that men and women are always called to direct their steps towards a truth, which transcends them. Sundered from that truth, individuals are at the mercy of caprice, and their state as person ends up being judged by pragmatic criteria based essentially upon experimental data, in the mistaken belief that technology must dominate all. It has happened therefore that reason, rather than voicing the human orientation towards truth, has wilted under the weight of so much knowledge and little by little has lost the capacity to lift its gaze to the heights, not daring to rise to the truth of being."

Preface

I have compiled ideas for this book since my discussions with Father Joe Hopkins, S.J. in the late 1960s with only a vague idea of what I was trying to accomplish: that a concept of God is foundational for the sciences, philosophy, anthropology, psychology, and sociology as well as theology and major religions. The concept of no god and atheism is fallacious; it is nonsense to deny the universality of God, which plays a significant role in the lives of more than three-quarters of the world's people.

Religion has been in authoritative decline in the West for a century. The religious quest was a search for truth and to frame individual and social values, but now defends its dogma. As science revealed the true complexity of the universe, life and the atom, many religions failed to adapt and simply demanded belief in past views of cosmology and life. Some reinforced their positions by demanding "deeper" belief while others simplified dogma altogether with messages like: "believe and receive God's bounty."

Modern physics and genetics do not reveal a simple universe. If we are to trust in the ultimate power of our God-given reason to discern new insights in our quest for truth, we cannot reduce God or his role in the universe to simple religious dogmas.

The book's title "Faith and Reason" addresses something that the world needs to hear about and understand. We have evolved from cells to organisms, to plants, to animals, to self-aware persons, to family, to tribe, to community, to state, to country, to global community and at each step a higher collective consciousness is created.

The need for consideration of a theological framework for religion and its purpose in society is presented in this book. Its purpose is twofold. First, it acts as a roadmap for understanding societal and personal improvement and advancement initiatives that use a holistic view of the history of relevant philosophical perspectives to present how belief and commitment to community can guide improving the human condition. Second, it presents detailed modeling constructs, and references to specific thinkers and philosophers to demonstrate the need for a spiritual relationship with God and each other.

An ecumenical theological structure, community and process are proposed to support the construction of an authoritative system of universal values. This system must help members of all religions understand the need for a spiritual relationship with God and each other. It must show how faith leads to rational ways that each of us can work together to reach our true potential and self-awareness. It allows us to use our God given consciousness to progress toward our ultimate purpose as individuals and members of society.

The requisite systems engineering discipline inherent in the particular graphical modeling approach will also assist the implementation of better religious practices. The underlying methodology uses a process modeling method, which has mainly been used for business process reengineering in the manufacturing industries. It is also founded on the Viable Systems Model, making it uniquely comprehensive. The scope of the models in the book is the communication and collaboration value chain, which captures teaching and mentoring throughout the life stage of people, communities and world societies.

The model considers formal and informal communication with a major focus on capturing change and individual advancement. This book synthesizes ideas from the major religions, philosophers and scientists, process models, Viable Systems Theory, and experience with improving people-capability in many societal and business contexts.

Social science treats religion as a universal phenomenon, but its concept of religion is very low, based on primitive rituals, mythology, and institutions. This book agrees with Eastern sages that the divine is equally present, but not equally manifest everywhere. Every human being is innately divine whose ultimate goal of life is to manifest his/her inherent divinity. Injuring another then, is a crime against his/her manifest divinity, the community and God.

Improving personal and community communication leads to better relationships and the expectation that we can – and should – depend on each other for improving our lives. Ultimately this will lead to - what is meant by Faith and Reason - enhanced confidence, individual courage and clarity, where and why we belong, and productive communication with each other.

This book represents the result of the struggle to find a working solution to the universality of God in the universe (the OED defines universality as the quality of being true in or appropriate for all situations). The chapters present God's authority within religions and the sciences through: individual transcendence, relationships in community, values and evolution of Western Society, evolution of intellectual disciplines, an adaptive ecumenical theological community, and a framework to answer questions about why we are here.

I would like to thank Ray and Winifred Adams, Mohammed Arif, Susan Clark, Rev. Jack Clark, Bill Cashman, Douglas Foran, Rev. LeRoy Haynes, Jacqueline Jones, Jeffrey Jones, Martin Maehr, Michael Sterling, and Glen and Kathryn Zerler for their critiques. I would like to express my deep appreciation to my wife Judy for her attempts to understand and improve the readability of this book while proof reading various revisions more than twenty times!

JimIJones@aol.com

Table of Contents

Outline 1

Belief *is required to begin reasoning* 2

Individual *transcendent observation can lead to revelation* 11

Community *validates revelation and integrates it into religion* .. 25

Religion *infused society with values derived from revelation* 40

Society *is supported and guided by religions* 56

Knowledge *is revealed faster than religions have adapted* 73

Theology *must guide religions to a universal value system* 90

Value *from religion is guided by an ecumenical theology* 106

> *The order of the world is no accident. There is nothing actual which could be actual without some measure of order. Religious insight is the grasp of this truth: That the order of the world, the depth of reality of the world, the value of the world in its whole and in its parts, the beauty of the world, the zest of life, the peace of life, and the mastery of evil, are all bound together - not accidentally, but by reason of this truth: that the universe exhibits a creativity with infinite freedom, and a realm of forms with infinite possibilities; but that this creativity and these forms are together impotent to achieve actuality apart from the completed ideal harmony, which is God.*
>   -Alfred North Whitehead, Religion in the Making

Faith and Reason Outline

Belief: Guide reason with belief	• Theologies • Panentheism • Imaginings • Atheism • Benefits	• Why a framework for religions and science? • How does one relate to God? • How are religions constructed? • Why imagine a panentheistic God? • How does A.N. Whitehead imagine God?
Individual: Be aware of transcendent self	• Perception • Consciousness • Process • Transcendence • Intuition/Revelation	• What is the nature of human spirituality? • What is individual consciousness, and perception? • How does human perception filter reality? • How does transcendence improve behavior and well-being? • How might God affect human perception?
Community: Validate Revelation	• Tribes • Guilds • Community Paradigms • Religious groups • Mores / Codes	• How did communities evolve? • What are the criteria for membership? • What historical role did religious communities play? • How can scientific communities provide a framework for a theological community? • What is a Societal Religious Community of Practice?
Religion: Worship in Community	• Images of God • Worship • Man's religions • Belief • Soul	• What are the elements of a religion? • How did religions evolve? • What are the major world religions? • What comprises a rational religion? • What role do sacred texts play in religion?
Society: Build a Secular Religious Society	• Social Imaginary • Economics • Political • Socio-cultural • Humanism • Religious Suppliers • Societal Ethics	• How does social imaginary differ from theory? • How do primitive, traditional and capitalistic societies differ? • How do cultural, political and economic systems interact? • How did Western Societies evolve from Christianity? • What are some modern forms of western religions? • How has the church emerged over centuries? • How must religion/churches evolve to influence society?
Knowledge: Seek Truth	• Disciplines • Science • Evolution • Atheism • Randomness	• What disciplines innumerate mankind's knowledge? • What does one know from science? • Why is no god not provable? • What does one know from religion? • How must religion relate to all disciplines?
Theology: Frame an Ecumenical Theology	• Philosophy • Quality / ideas • Quantum Theory • Berkeley, Pirsig, James, Northrop, Whitehead	• What is reality? • What is static and dynamic quality/value/knowledge? • What are levels of consciousness? • What are the stages of science? • What are the elements of value? • What might the mind of God be like?
Value: Guide Human Behavior	• 21st Century Theology • Theology Paradigm • Purpose • Value/Ethics	• How should religions body of knowledge be extended? • By what ethics will people live their lives? • What is our purpose: vocational, familial, societal, spiritual? • What will a science of religion look like and do? • How can orthodox churches make change?

Belief

Belief is required to begin reasoning.

This book's thesis is that belief in God is necessary to build a viable framework for religion and science. Its intent is to revive the religious quest to understand God's universality and re-establish the authority of religion.

Belief in a science is misplaced. Science explains physical phenomenon within limits; it is not 'truth.' The scientific method is used to validate theories with facts and experiments. However, human behavior is driven by value not fact.

Nevertheless, Harvard Professor Alfred North Whitehead argued that Christianity was essential for the rise of science: "It must come from the medieval insistence on the rationality of God, …the search into nature could only result in the vindication of the faith in rationality." His panentheistic concept of God provides a framework for a theology that synthesizes all knowledge and provides an ecumenical interpretation of the major world religions.

Rigid adherence to dogma is making religion irrelevant in the public square. Among other things, an increase in atheism has emerged. To bring back the confidence in religions, they must return to the reasoned faith that initiated science in the first place.

In the last decade, new atheists have attacked orthodox religions and a supernatural view of God, mistakenly 'proving' that God does not exist while rehashing 300-year-old arguments. They do not account for Whitehead's view of God in *Process and Reality*. A renaissance of theological thought comparable to Darwin's and Einstein's theories has arrived that seeks to restore the authority of religion and God in a way that embraces the sciences, but atheists fail to recognize or acknowledge it.

Eastern gurus assert that science, psychology, sociology, ethics, aesthetics and human consciousness fall under the umbrella of philosophy and that such a philosophy must embrace a concept of God. Alfred North Whitehead's[1] Theory of Organism provides such a foundation while science alone does not. His panentheism creates a narrative about God that combines concepts of both Eastern and Western philosophy. Panentheism understands God and the world to be interrelated with the world being in God and God being in the world.

Whitehead's panentheism is augmented by Eastern philosophy as revealed in their sacred text: Vedanta.[2] Panentheism's concept of God has been used by theologians to develop a rational view for all major world religions and provides a theological framework to guide religions toward a set of universal human values.

Since the dawn of time people observed: "What is" and asked "What if?" It starts with practical applications like fire, weapons, wheels, and buildings, but coincidently people wonder what causes pestilence, disease, well-being, wealth, poverty, weather, seasons, night and day, sun and stars. Unknown causes were attributed to God. Not surprisingly, individuals desired to know the "Mind of God." Having come to some revelation about various causes, they wished to communicate it to their fellows along with proposed behavior about how to elicit a favorable judgment from God with subsequent benefits.

---

[1] Whitehead taught mathematics at Trinity College Cambridge, physics and philosophy at University College and Imperial College London and philosophy at Harvard University.

[2] Vedanta includes elements of Buddhism and Jainism. It existed prior to but became prominent, with its most influential advocates: Shankara 8th century CE, Ramanuja 11th century CE and Madhva 12th century CE. Vedanta's fundament teaching is humans are innately but not equally manifest in the Divine.
In John 8:12, Jesus said: he that follows me shall not walk in darkness, but shall have the light of life.

Religion is built on: ritual (habitual performance of a system of actions with defined benefits), emotion, belief, and rationalization (adjustment of beliefs into a system of principles that guide us and how we relate to others). First there is ritual, then emotion leads to commitment and then rituals are followed for the emotion, which they generate. Belief appears to explain ritual and emotion. Rationalization and belief established the great religious conceptions.

Collective rituals and emotions bind savage tribes; they represent the spirit raised beyond the task of supplying animal necessities. Conversely, religion in decay sinks back into sociability. Just as ritual encouraged emotion beyond the mere response to necessities, rational religion goes beyond thoughts of immediate circumstances and creates wonder about God and the universe.

Whitehead in his essay, *Religion in the Making*, defines Rational Religion as: religion whose beliefs and rituals have been reorganized with the aim of making it the central element in a coherent ordering of life - an ordering which shall be coherent in respect to the elucidation of thought, and the direction of conduct towards a unified purpose commanding ethical approval.

Religious dogma attempted to formulate in precise terms the truths disclosed in the religious experience of mankind. In a similar way, the dogmas of physical science attempt to formulate in precise terms the truths disclosed in the sense perception of mankind. Religion is founded on the value of: 1] an individual for oneself; 2] the diverse individuals of the world for each other; 3] the objective world, which is a community derived from interrelations and the existence of its individuals.

Mankind deludes itself into believing that it knows things with certainty and then proves that it does not. Atheists speak derisively of the god of the gaps, failing to recognize that the gaps in our ignorance are infinite. Gandhi said: "God is, even though the whole world denies him."

Man's attempt to imagine God as a force in the environment through religion has been the foundation upon which Western knowledge has progressed. In the last several hundred years, orthodox religions have separated themselves from this search for knowledge and now simply defend their dogma.

Whitehead's panentheism argues against atheism and agnosticism. Panentheism is not used to argue against theism. In his book, *Panentheism: The Other God of the Philosophers,* John Cooper gives a detailed account of the history of panentheism and then argues against it as a Christian theologian.[3]

However, Whitehead's panentheism provides a framework for a philosophy of science and an ecumenical interpretation of the major world religions. Whitehead's panentheism was continued in Hartshorne's process theology where without God, the world would be a static, unchanging existence radically different from the actual world of experience. An eternal and temporal God provides possibilities to change and develop the world.

---

[3] Evidence of panentheism was present in Ikhnaton (1358 BCE), the monotheist Egyptian pharaoh; in the Upanishads; in Lao-Tsu (4th century BCE); in Judeo-Christian scriptures; in Plato's (400 BCE) concept of Forms and the World; in Plotinus' (270 CE) identification of God with the world; in Proclus (485 CE); and in Pseudo-Dionysus (6th century) drawing upon Plotinus.

In the Middle Ages, evidence of panentheism was present in Eriugena (877), Eckhart (1328), Nicholas of Cusa (1464), Boehme (1624), Bruno (1600) and Spinoza (1677). Later 17th century thinkers such as Edwards (1758), and Schleiermacher (1834) thought of the world as a development from God.

Karl Krause (1832) created the term 'panentheism.' Hegel (1800) and Schelling (1831) sought to unify reality by means of dialectic thought culminated in process philosophy's where God is affected by the world.

As a world-renowned spiritual and religious figure Mahatma Gandhi developed civil disobedience as the way to rid India of British rule. Martin Luther King adopted his methods to end segregation in the south. Gandhi's life and politics were greatly influenced by his sage, Shrimad Rajchandra. In response to friends' efforts to convert him to Christianity, Gandhi asked fundamental questions on Hinduism of Rajchandra whose replies in Gandhi's words "were so logical, appealing and convincing that I regained my faith in Hinduism." Although Rajchandra was well versed in many Hindu and Western religious sacred texts, he was partial to the Vedanta as was Gandhi. Whitehead's panentheism is similar to the Vedanta's view of God.

Throughout the book reference is made to doctrines of the Vedanta school, which is one of six schools of philosophy within the four sects of Hinduism. The Vedanta contains 108+ Upanishads (inner or mystic teaching). Its most fundamental teaching is that all exists in the divine.

The intent of this book is to revive the religious quest to understand God's immanence and re-establish the authority of religion. This is presented in seven areas (quotes are from Gandhi):

1] Individual: "Each one prays to God according to his own light."

2] Community: "The best way to find yourself is to lose yourself in the service of others."

3] Religion: "The essence of all religions is one. Only their approaches are different."
"I like your Christ, I do not like your Christians. Your Christians are so unlike your Christ."

4] Society: "Human society is not divided into watertight compartments called social, political and religious. All act and react upon one another."

5] Knowledge: "There is an unalterable law governing everything and every being that exists or lives."

6] Value: "Truth stands, even if there be no public support. It is self-sustained."

7] Theology: "Before the throne of the Almighty, man will be judged not by his acts but by his intentions."

Individual transcendent observation can lead to revelation. Seeing things in a new way can be revealed as an extrapolation of a mathematical formulation or scientific theory. However, most new knowledge is revealed by transcendent intuition – seeing things in ways that no one has observed before. Newton's Laws, Darwin's Theory, Confucius' Analects, Christ's parables, art, literature, engineering, mathematics, architecture, technology, and science are evidence of a continuous stream of revelation. All explicitly defined knowledge (scientific or religious) either comes from the mind of an individual or is revealed to an individual by God. Eastern sages argue that Truth is revealed only through intuition in transcendent meditation within God. Further, individual spiritual transcendence is proposed as an approach toward providing emotional wellbeing, virtuous behavior and novel ideas.

Community validates revelation and integrates it into religion. Communities are populated by individuals who have acquired and used knowledge and skills to support critical areas of society and also serve to critique the novel ideas of its members. Scientific revelations (theories) are validated by experiments, predicted observations, and/or confirmation of mathematical derivations. In a scientific or engineering community most revelation is viewed skeptically, especially if it invalidates established theories (dogma). The Physics community ostracized Albert Einstein until his Theory of Relativity was validated by observation of cosmological phenomenon. Religious revelation must also be viewed skeptically until a theological community validates it.

Religion infused society with values derived from revelation. Religious validation has tended toward extremes: all new ideas are summarily rejected if they contradict sacred texts; conversely, new age ideas from a charismatic leader are accepted without question. Religious validation is key to evolving systems of worship and values that respond to a rapidly changing world order (population, technological advances, ecological degeneration, etc.). Religion imagines God as that force which created and sustains the universe and speculates about questions like: How did the universe get here? What is the meaning of life? What happens after death? What does God want? Because religions are based on man's fallible, divinely inspired, imagined view of God, sacred texts represent extraordinary diversity in beliefs.

Society is supported and guided by religions. Society is composed of communities that support the military, commerce, government and disciplines with socio-cultural norms for governance and citizen welfare. Societies evolve from their religious/cultural, economic and political systems. Whitehead asserts rational religion lies at the foundation of Western Society. Western society evolved from a theocracy to a secular democracy, which is being subjected to the detrimental systematic elimination of religion in the public sphere. Some would argue that proper behavior is innate in humans and that a society can provide a quality of life for its citizens without God. *This book agrees with the Roman Emperor Constantine and the USA's founding fathers: Christianity has and still provides a foundation for proper behavior and human values in Western society.*

Knowledge is revealed faster than religions have adapted. Communication technology, mathematics, engineering and the scientific method are translating theories into innovations at a spectacular rate. With the Internet and various other technological advances, a world community has been created with world-wide instant communication.

In the past, religion was integral with many bodies of knowledge such as: science, art, literature, philosophy, music, education, medicine, government, policy and law. Some would say religion is now out of touch with modern reality. Since people's lives are value driven, religions must embrace a changing world with a universal concept of God and not retreat into a 'personal' view of God.

Theology must guide religions to a universal value system. Human life, institutions and all disciplines are driven by value, which science excludes. Whitehead's Theory of Organism provides a foundation for bringing philosophy, physical and social sciences, mathematics and religion under one framework. God's primordial and consequent nature elegantly explains reality and has been adopted by theologians of all world religions. Two thousand years ago theology was a continuously evolving search for God's truth, to explain everything. Religions must return to that tradition within a collaborative religious environment to: improve human welfare, encourage harmonious beliefs of disparate religious views and provide collective and individual guidance to people.

Value from religion is guided by an ecumenical theology. Aided by Whitehead and others, this book attempts to present a limited view of God's truth and renew the theological quest for truth. It contends that God interacts with humankind through natural laws (e.g. Biology, Physics, Sociology). The book proposes that theology must evolve to provide a system and discipline that provides moral guidance to our present humanistic secular western society or risk falling back to 16th century theocracy. To prove God exists, philosophy and science have proved inadequate (e.g. refutations of Augustine, Aquinas, Descartes, Kant 'proofs of God's existence'). What emerged is the synthesis proposed earlier as a narrative explaining how and why religion:
1] Benefits Individual well-being,
2] Precipitates and validates revelation,

3] Was used to charge societies with value,
4] Provided the foundation for science and belief in rational thinking,
5] Can evolve based on an ecumenical theology that guides all religions and derives ethics and values for individuals, communities and societies.

Author's Reflection

Faith in no god requires a great deal of belief absent reason. Atheists' arguments fall into several categories: science can determine value, awful things are done in god's name, belief in god is a silly artifact left over from primitive savages' superstitions, and neo-Darwinism "proves" there is no god.

The book argues that these assertions are not viable in the light of what is now known about biology, the universe and Alfred North Whitehead's panentheism. According to Cooper, some modern philosophers align panentheism with science and include: Calvinist American preacher, Jonathan Edwards; Catholic Jesuit priest, Pierre Teilhard de Chardin; Christian theologians John Cobb and David Griffin; Professor Paul Tillich's theology. Little of this thought finds its way into modern religions.

Many people and ministers argue that religion is about belief not reason. Not so, St. Augustine (3rd century CE) and St. Thomas Aquinas (12th century CE) both tried to align Christianity with Greek science. *One purpose of this book is to acquaint readers with the importance of scientific methods in rationalizing religion for it to be a positive force in society.*

Individual

Individual transcendent observation can lead to revelation.	
• Perception • Consciousness • Process • Transcendence • Intuition • Revelation	*What is the nature of human spirituality?* *What is individual consciousness, and perception?* *How does human perception filter reality?* *How does transcendence improve behavior and well-being?* *How might God affect human perception?*
Seeing things in a new way can be revealed as an extrapolation of a mathematical formulation or scientific theory. However, most new knowledge is revealed by intuition – seeing things in ways that no one has observed before. Newton's Laws, Darwin's Theory, Confucius' Analects, Christ's parables, art, literature, engineering, mathematics, architecture, technology, and science are evidence of a continuous stream of revelation. All explicitly defined knowledge (scientific or religious) either comes from the mind of an individual or is revealed to an individual by God. Eastern sages argue that Truth is revealed only through intuition in transcendent meditation within God. Further, individual spiritual transcendence is proposed as an approach toward providing emotional wellbeing, virtuous behavior and novel ideas.	

Individuals seek to understand life, nature, earth, galaxies and the universe; they seek to understand or deny God in light of mankind's existing knowledge and transcendent truth. Who am I? How will I behave? What do I want? What will I learn? How will I contribute?

Science requires collecting and synthesizing facts relevant to a problem into a coherent theory, and testing it. The scientific method synthesized relevant facts into Newton's physics, Einstein's cosmos or Darwin's Evolution.

For three centuries, scientific and rational methods have dominated Western philosophy. However, in *Adventures in Ideas*, Alfred North Whitehead contends the focus solely on intellect denies conscious access to the more fundamental kinds of meaning that rational thought can structure, analyze, and critique, but cannot engender. He writes: "Knowledge is always accompanied with accessories of emotion and purpose… To understand is always to exclude a background of intellectual incoherence. But Wisdom is persistent pursuit of the deeper understanding, ever confronting intellectual systems with the importance of its omissions."

Indian philosophy claims that intuition, not reason, is the true measure of Reality that introduces an individual to Truth. Western science has progressed through reason and experimentation, but leaps forward through intuition. However, Reality as understood in oneness with God, still must be validated using reason and experimentation. *Furthermore, if Truth from intuition is to be 'useful,' the individual must be thoroughly versed in the discipline in which he hopes to intuit new truths. In St. Augustine's words: "God does not speak by these things, but by the truth itself—if anyone is prepared to hear with the mind rather than with the body."*

Western philosophy evolved from Descartes believing in a reality of objects observed by a subject (i.e., us). From this beginning, a Subject/Object Metaphysics (SOM)[4] dominates philosophy's view of reality: objects that can be seen and touched exist independently of the subject. SOM does not reconcile with what is known about quantum particles. Hindu philosophy views reality and its observer as one and the same entity and claims that in SOM analysis truth is distorted and falsified, because existence gets separated into subject and object.

---

[4] Metaphysics is concerned with explaining the fundamental nature of being and the world by which people understand the world: existence, reality, objects, space, time, causality, and probability.

Whitehead's view of process as the fundamental unit of reality evolved from William James' denial that the subject-object relation in metaphysics is fundamental to knowledge. Alexander (d.1938), Bergson (d.1941), and Morgan (d.1936) introduced divine development into conceptualizing physical reality and provided background for Whitehead's panentheism.

In the *Theory of Organism*, Whitehead provides a Western scientific definition of this Hindu view. He proposes that the only reality is process, that everything has a form of intelligence and that objects change moment to moment. One may perceive a rock as the ultimate fundamental unchanging object, but at the quantum level it changes every nanosecond (it keeps its shape by obeying physical laws). Plants, organisms, animals exhibit progressively higher orders of consciousness culminating in humans' god-like consciousness. One sees projections of reality, considerably altered by one's fields of consciousness. The question to be answered is: how does one transcend a limited perception of reality in an attempt to acquire new insights (e.g. scientific or religious revelation).

Whitehead sees William James as the initiator of a qualitatively new mode of thought emerging from Descartes' dominant Cartesian philosophy: "Neither philosopher finished an epoch by a final solution of a problem. Their great merit is ...(that) each of them open an epoch by their clear formulation of terms in which thought could profitably express itself at particular stages of knowledge, one for the 17th century, the other for the 20th century."

According to William James, An individual's spirituality may be judged objectively by its impact on behavior. Spiritual transcendence frames how humans experience the divine and might provide a dimension necessary for guiding 'proper' behavior and emotional wellbeing.

In his book, *Talks to Teachers on Psychology*, James proposes that mental processes be evaluated by effects on practical life; those functions of mind that do not refer directly to the world's environment – ethical utopias, aesthetic visions, insights into eternal truth, logic – could not be carried on at all by a human individual, unless the mind that produced them in him were also able to produce more practically useful products. Here are definitions abridged from James' book:

Stream of consciousness *goes on when one is awake; it is a succession of states, or waves, which constantly pass and repass that constitute inner life, thoughts and emotions.*
Field of consciousness *transforms streams of consciousness; it contains physical sensations in our bodies and the objects around us, memories of past experiences and thoughts of distant things, feelings of satisfaction and dissatisfaction, desires and aversions, and other emotions.*
Apperception *takes fields of consciousness into the mind. Every impression that comes from without (hear, view, smell) enters our consciousness only to be connected with memories, interests and ideas already there to produce a reaction. It is disposed of as ideas that designate inner objects of contemplation (e.g., things: 'Caesar', 'sun' or classes: 'animal kingdom' or abstractions: 'rationality').*
Conscious process *is derived from apperceptions and pass over into open or concealed motion. A simple case is a mind possessed by only a single idea that connects to an impulse to immediately discharge. But now suppose two ideas are in the mind: A taken alone would discharge in certain action, but B suggests inaction. Idea **B** may inhibit **A** unless I consider it my duty to perform **A** despite **B**.*
Moral acts, *in their simplest and most elementary forms, consist of an effort of conscious process by which an idea is held fast, which but for the effort of attention, would be driven out of the mind by other psychological tendencies present. Moral acts can inhibit certain detrimental behaviors and be considered a form of virtue.*
Inhibition by repression *in moral acts moderate behavior: both the impulsive idea and the idea that negates it, remain together in consciousness, producing an inward tension.*
Inhibition by substitution *in moral acts also moderate behavior: the inhibiting idea supersedes the idea to prevent it. Substitution is usually better to employ than repression, and is the key to virtue.*

How is *inhibition by substitution* best achieved? Meditation, found in all religious systems, claims to moderate behavior toward moral acts. Abrahamic religions' prayer is a form of meditation; Buddhists recite mantras in meditation to tranquilize the mind to a state of receptivity. Meditation's goal is to achieve a particular psychic result. Christians who see Jesus, or Buddhists who converse with Buddha, may be satisfied that their meditative purpose is fulfilled.

The Hindu sage knows that false apperception is always gained through: eyes, ears, nose, tongue, touch, mind and dreams. Mandukya Upanishad proposes a method for achieving Truth through mediation that transcends one's awareness from waking to dream to deep sleep to ultimate consciousness in oneness with the Absolute (God). Know it as 'That which is' said Saint Augustine.

Unwholesome impulses come from an unwholesome mind, and as it becomes purified, healed of its disorders, worldly cravings cease to accumulate. The sage abhors wrong action (*inhibition by repression*) and takes greater delight in deeds that are rooted in generosity, benevolence and wisdom (*inhibition by substitution*).

In contrast, within Gnostic Christianity one comes to God through knowledge of self, knowledge of the world and knowledge of God through meditation. Knowledge of God is at best unfathomable. Knowledge of oneself comes from proper habits,[5] apperception, and reflection

---

[5] **Habits**, practical, emotional, and intellectual, systematically organized for our weal or woe bear us toward our destiny. Our activity is 99% habitual. Education organizes acquired habits.

"Proper" habits are used by a virtuous being to manage his stream of consciousness; when habits conflict; the least tense virtuous one uses inhibition by substitution: habitually tell the truth, not to avoid the evil of lying, as to arouse enthusiasm for honor and veracity.

Skill (proper habit) is learned capacity to carry out pre-determined results often with the minimum outlay of time and/or energy.

on one's experiences. Knowledge of the world comes from our relationships in community and society and participation in disciplines.

In addition, Quantum Theory and General Relativity confirm the Vedanta's Truth that phenomena known through cognition (and senses) do not correspond to the quantum world or the cosmos. In his book *Process and Reality,* Whitehead builds on these theories by proposing that all reality is a dynamic stream of consciousness.

Descartes Subject/Object Metaphysics (SOM) assumed our senses are the only reliable reality. Whitehead thinks otherwise. He redefines James' field of consciousness, apperception and conscious process, into three distinct stages. Whitehead provides a detailed view of how one's mind processes external stimuli as constituted in the past (causal efficacy), as taken up into the present (presentational immediacy) and as reference to immediate future (symbolic reference). Since the two primitive modes of perception are incapable of error, symbolic reference (human conscious process) introduces this possibility.[6]

Whitehead's ultimate principle of reality is a process event called an *occasion*.[7] Reality is never static and is constantly changing. Creativity does not exist in any other

---

[6] **Causal Efficacy** is represented by fields of consciousness, where human beings are unconsciously influenced by objects in their past before they perceive something with their senses.

Presentational Immediacy is the apperception with one's senses to vaguely perceive the external environment, interpreted by past objects, and projects that external data into symbols of the mind.

Symbolic Reference is the conscious process whereby symbols transition into meaning when the interaction of perception in the modes of causal efficacy and presentational immediacy elicit consciousness, beliefs, emotions, and usages. No one directly perceives external reality.

[7] **Occasions** apply not just to human apperception, but to animals, plants, rocks, etc. and at levels of reality from atom to cosmos.

form than occasions. When an occasion achieves its result, it ceases to be an occasion and becomes history, eternally unchanging in the form it has taken.

Finally, Whitehead defines God as imminent and primordial in the universe and the purveyor of all knowledge in the form of eternal objects (ingredients that make up real/actual entities and objects). Humans interpret this knowledge in various ways: e.g. physics, genetics, Hamlet, music, etc…

Whitehead says that creativity is *novelty of instance*. Novelty of instance is a new result derived from previously actualized old data; *novelty of kind* introduces new data into the stream of process. God influences novelty of instance in all things identified by 'descriptive words' such as 'yellow' and 'car.' There are an infinite number of these eternal objects.[8]

Eternal objects 'exist' as potential but are activated only when an occasion realizes a particular combination of objects. An observer prehends in a mode of causal efficacy to anticipate the appearance of "a yellow car" in his stream of consciousness. Upon perceiving a vehicle in the mode of presentational immediacy, he limits occasion with his subjective aim[9]: model and color.

---

[8] **Plato's Theory of Forms** asserts that non-material abstract forms (or ideas), and not the material world of change known to us from sensation, possess the highest and most fundamental reality. To Plato, Forms are the only true objects of study that can provide genuine knowledge. Whitehead's eternal objects are the application to process reality of Plato's Forms.

[9] **Subjective Aim**: Most prehensions (concrete modes of analysis) of the world focus on the past. They are the feeling and analysis of the entire world for that occasion, but they do not complete that occasion by themselves. From prehending data and admittance of new possibilities comes a unified ideal for the end result. This is the subjective aim--a projected concrete form in which to resolve the diversity of feelings of the primary phases of the process.

The 'satisfaction'[10] of the subjective aim in the mode of symbolic reference imposed upon the prehensions is: the yellow car. A broader aim might include brand and style.

For example, consider two different projections of the same reality: husband and wife walk together in the woods. The man walks for contemplation and fitness and his wife walks to be with her "plant and wildlife friends." The husband walks briskly for aerobic fitness while contemplating transcendence. His wife wanders while contemplating trees, birds, flora and fauna. Their subjective aims and fields of consciousness are vastly different. The husband limits his perception of external reality while directing his focus internally. The wife limits her internal focus and maximizes her external awareness of the natural environment. Note that the wife may interrupt her husband's field of consciousness by pointing to an eagle in a treetop while the husband may interrupt his wife's external awareness by urging her to walk faster or launching into an arcane description of transcendence.

Both husband and wife are immersed in the same reality, but will recall it with very different perspectives. He will tell you how many miles they walked and relate some aphorism that he conjured up during the walk. She will tell you about eagles, dwarf iris, trees, trash, road texture, and wetlands. Both experienced the same reality in space and time, but it resulted in different 'Occasions' for both walkers who used a different field of consciousness and subjective aim to project their internal reality.

Subjective aim affects the value of experience differently. The husband's perception is severely restricted because

---

[10] **Satisfaction** achieves the unity proposed in the subjective aim. The process is finished - all felt aspects have been reconciled in a unity of feeling or 'negative prehension'- denial of access into the satisfaction. With the satisfaction, the occasion is 'done'- fixed form of resultant unity. The satisfied occasion loses its actuality as it passes into history as fixed data.

his field of consciousness is directed inward toward an abstract problem requiring symbolic reference to develop his own approximation of God's eternal object. His novelty of instance may be to discover a new way to explain Whitehead's view of perception.

The wife's subjective aim was to reduce her prehension in the mode of causal efficacy so that she could experience more in the mode of presentational immediacy in her natural environment. Her use of symbolic reference is to categorize the various species that she encounters. Her novelty of instance may be to identify an endangered plant not known to the area.

The point of these initial explanations is to begin to frame the issues associated with the past, present and future purposes of religious systems. Finding yellow cars and walking in the woods are simple examples to illustrate concepts of perception and thought. This view of perception will be important in arguments presented in later chapters where communities reinforce their wrongly believed perceived projections of reality.

The infinity of the eternal objects to be considered represents God's primordial nature. In His consequent nature, God participates in harmonizing physical and conceptual prehensions[11], and the subjective aim of every occasion. God is not merely a warehouse of forms of possibility, God's consequent nature conceptualizes all possibilities. Whitehead calls these 'valuations'.

---

[11] Whitehead describes his system of speculative philosophy as a 'philosophy of organism,' in that reality consists of interrelated and mutually dependent parts that sustain vital processes. Final realities that make up the world are 'actual occasions,' or 'actual entities.' These are the concrete facts on which our thoughts and feelings are based. The other basic elements of human experience include 'prehensions' (concrete concepts and feelings) of actual entities, and the 'nexus' (or system of relationships), which connects the development and functioning of all actual occasions.

Value is tied to Occasion through the harmonizing of prehensions (God's consequent nature) with eternal objects (God's primordial nature). The aspects of value are the building blocks for individual and societal ethics.

A nexus threads a group of occasions together into a society of occasions in Whitehead's view. Writing this book was a nexus of occasions. The satisfaction – yellow car – does not take much harmonizing, but writing a book about God requires a great deal of harmonizing of eternal objects and prehensions in a nexus of occasions.

The author's efforts to preserve his own reality in the mode of symbolic reference are reflected in the nexus of satisfied occasions documented by this book. The unifying principle of the nexus is a rational view of God. The book is now history eternally unchanging in its form as part of God's primordial nature. God's consequent nature and harmonization are reflected in whatever truth the book may contain. Its falsities and inadequacies result from the author's ignorance. The book may add to a reader's mode of symbolic reference as he prehends its ideas, which may then change his use of symbolic reference as he processes new occasions.

Hindu meditation changes consciousness, thereby achieving transcendence and experiencing revelation (not hallucination). This may prove to be difficult especially for a scientifically trained Western mind. However, many a researcher has studied a problem and the associated facts for months without formulating a relevant theory, only to wake up with an answer through unconscious revelation – the "Eureka!" moment. Ask someone who has provided a new solution to a problem and as often as not you will get a shoulder shrug for an answer. Prayer may achieve transcendence but only if one stops asking and listens: "Be still and know that I am God!" Psalm 46:10.

In Whitehead's terms, meditation suspends 'causal efficacy' and 'symbolic reference' (emptying the mind) so that more of the external reality and God's eternal objects are projected onto one's consciousness, and God's harmonious nature produces novel occasions and expands one's symbolic reference.

Moral acts require the resolution (symbolic reference) of a conflicting system of ideas (ignorance of eternal objects) to avoid evil and act with honor and veracity (subjective aim). Unaided by transcendence, moral acts may require a significant number of inhibitions by repression resulting in a sense of loss and feelings of deprivation.

By being at one with the universe through meditation or prayer, God's consequent nature can provide an independent and detached view of self and reduce potential errors of symbolic reference. It can help quiet a relentless stream of consciousness that makes us human and provide the proper habits of mind that can produce truth and virtue. Value in the form of personal virtue and societal ethics (occasions harmonized by God's eternal objects) will be discussed in subsequent chapters.

Author's Reflection

Regrettably, some deny intuition in science and spiritually in life, and sadly, remove themselves so far from reality to conclude: there is no god. Atheists delude themselves with the idea that random events and natural selection explain the universe, planets, life, humanity and no god.

Randomness is a valued statistical tool, but it is delusion to use it as an explanation of cause; it is used when one has no idea about what is happening; it supports statistical analysis to better understand clusters of data.

Whitehead provides a foundation for a rational view of religion, but his book, *Process and Reality*, is considered as the most opaque philosophy ever written. Again, he asserts: apperception occurs in three overlapping processes. The first is our field of consciousness that filters how we see something (causal efficacy). The filtered reality we perceive is represented by a subset of what is there (presentational immediacy). Then our conscious thought (symbolic reference) transforms our filtered perceptions into a final product (occasion) in our mind. Whitehead contends the only reality is process. He has interpreted our perception in terms of the never-ending motion of quantum mechanics.

For example, in the woods, we expect to see various plants; so when a tall brown object with shaggy green things at the top presents itself to our vision, from the categories of information already anticipated by our mind, we see a tree. Whitehead would point out that in addition to the stream of consciousness as a filter, our sense limitations represent a filter, which allows us to identify this as a tree. However, in a quantum spectrum of reality, atoms moving at a spectacular rate would represent the tree. We do not observe this 'reality' without equipment, mathematics and scientific theories.

When Newton created the laws of physics to describe the motion of our solar system, he already had data and theories stored in his brain, which allowed him to draw his conclusions, but they were not obvious from past data and theory. Revelation occurs all the time in science, but it goes unrecognized as revelation because we think that it is logically deduced from rational thought; it is not.

To precipitate revelation, we consider changing our perception filter, so that we can look at the same environment and see a vastly different "reality." Transcendence is a process of changing how we perceive things allowing us to create novel ideas, which may or may not be a better approximation of reality. This not only applies to science, but also to literature, ethics, economics, behavior and religion.

Meditation suggests that we empty the mind, which would mean reducing the impact of both causal efficacy and symbolic reference in our perception of the reality that is before us. This may allow us to see a 'different reality.'

Hindu meditation claims to deliver us from delusion. It claims significant benefits: loving kindness, compassion, sympathetic joy, and detachment, which in turn helps us avoid wrong action and delight in right action.

I love my stream of consciousness and don't like to suspend it. Thinking about everything all at once all the time can be a significant advantage, when trying to solve complex problems. One might ask, if one has a wide and rapid stream of consciousness, why bother to suspend it?

Some of the time after surveying vast quantities of data, interviewing 20 or 30 people, summarizing observations, and then subsequently failing to find an adequate solution, I just go to sleep and wake up with an answer. Over the years, "magic" was not satisfactory to explain a solution to scientists and engineers or for being given money to proceed. I must show why my magical theory or solution is a consequence of observing and synthesizing real data. Transcendence goes unacknowledged.

Whitehead defines God as immanent and primordial in the universe and the purveyor of all knowledge in the form of eternal objects (much like Plato's Forms), which exist prior to all perception of reality. We as humans approximate these eternal objects into disciplines like physics, genetics, literature, music, etc. Whitehead further contends novel ideas result from intuiting reality and the harmonizing influence of God (magic). In a walk in the woods, the wife's and husband's field of consciousness are entirely different. They produce vastly different results (occasions) based on their stream of consciousness and intellectual processes that they apply to it.

Upon experiencing a revelation great or small, a person must validate whether or not it is brilliant or delusional. Unless he can convince people that God told him in a state of transcendent consciousness, a community of people must validate it. The scientific method requires that this revelation be put into some kind of theory, along with experiments that are capable of disproving the theory. The community then performs those experiments to disprove the theory. If they can't disprove it, the revelation is accepted as an approximation to the truth.

However, in religion and ethics, it is not always possible to develop a theory and experiments. It then becomes the job of the community, to determine the plausibility of the revelation. This requires selecting people with the experience necessary to critique the revelation. To answer these questions in the next chapter, we will look at how communities have evolved, what were the skills and experience of its members and how was the framework developed to support the community and its intent.

Science has had a significant impact on our view of truth and religion. Most religions have failed to accommodate these new interpretations of God's eternal objects. They have failed to see God's revelations that expand our perception of our world and universe; and hence, have failed the faithful who follow their religious traditions. They ask their followers to believe too much absent reason and have minimized the impact of their great religious traditions on the societies of the world. The remainder of this book points the way to a more equitable accommodation of religions and science.

Community

Community validates revelation and integrates it into religion.	
• Tribes • Guilds • Community Paradigms • Religious groups • Mores / Codes	• How did communities evolve? • What are the criteria for membership? • What historical role did religious communities play? • How can scientific communities provide a framework for a theological community? • What is a Societal Religious Community of Practice?
Communities are populated by individuals who have acquired and used knowledge and skills to support critical areas of society and also serve to critique the novel ideas of its members. Scientific revelations (theories) are validated by experiments, predicted observations, and/or confirmation of mathematical derivations. In a scientific or engineering community most revelation is viewed skeptically, especially if it invalidates established theories (dogma). The Physics community ostracized Albert Einstein until his Theory of Relativity was validated by observation of cosmological phenomenon. Religious revelation must also be assumed to be nonsense until a theological community validates it.	

Communities have supported critical areas of society. What is their purpose? What are the criteria for membership into the various communities? How do they operate? What purpose should theological communities serve?

Primitive religious communities evolved into ancient and modern guilds and scientific communities. Religions must establish theological communities that seek truth rather than simply demanding belief.

Previous chapters examined systemic revelation in religious systems. Now we turn to how this relates to groups of people, communities and society. As discussed previously, this will involve a new ecumenical theology that engages a critical theological community.

Individual transcendence is important in developing a personal spirituality and theology, but if one sees his revelation as a gift from God to be delivered to the masses, how is transcendent revelation separated from hallucinogenic nonsense?

Consider that the Old Testament Bible was passed on orally and was written and rewritten over hundreds of years in a community of priests and prophets. The final rework is thought have been written around 600 BCE in a collaborative group called the Deuteronomists.

Before considering a new kind of theological community, consider how modern day scientific and engineering communities and guilds evolved.

A guild (ref: www.masterartisan.com) is an association of people of the same trade or pursuits (with a similar skill or craft), formed to protect mutual interests and maintain standards of workmanship and ethical conduct. A guild is a trade union of sorts, since each crafter was a self-employed individual artisan or part of a small craft shop.

Regulated professions were a feature of the ancient and classical world. The Code of Hammurabi (sixth king of Babylon and a stonemason who died in 1750 BCE) specified a death penalty for builders, or masons, whose buildings fell on the inhabitants. Those in a position of special knowledge or trust were held accountable to the public for their advice and services.

Guild associations called collegia existed in ancient Rome before the 4th century CE. They were sanctioned by the central government and subject to the authority of the magistrates. When the Roman Empire fell, guilds disappeared from Europe for more than six centuries.

Guilds (ref: www.britannica.com) became possible in Europe with the appearance and growth of towns in the tenth century. Craft guilds arose soon after merchant guilds. Craftsmen banded together to regulate competition among themselves. They agreed on basic rules governing their trade, and set quality standards to form the first craft guilds. Guilds shaped labor, production and trade; they had strong controls over instructional capital, and concepts of a lifetime progression from apprentice to craftsman, journeyman, master and grandmaster. All guild structures were similar: a governing body with a leader and deputies, assistants and the members' assembly.

Islamic civilization extended this to the warraqeen, "those who work with paper." Early Muslims were heavily engaged in translating and absorbing all knowledge from all other known civilizations. Critically analyzing, accepting, rejecting, improving and codifying knowledge from other cultures became a key activity, and a knowledge industry began to evolve. From the 9th to the 15th century, warraqeen were engaged in paper-making, bookselling, and taking the dictation of authors, to whom they were obliged to pay royalties on works. A new work was publicly presented in a mosque to scholars and students; a high degree of professional respect was required to ensure that others did not simply make and sell bogus copies.

Modern guilds exist in different forms. European guilds have had a revival for craftsmen. Screen Actors and Writers Guilds exercise strong control in Hollywood because a system of intellectual property rights exists. Real estate brokerages display guild behavior: strong affiliation among all practitioners, self-regulation, strong cultural identity, little price variation, and traditional methods in use by all practitioners. US Law practice operates as guilds: every state maintains its own Bar Association, supervised by the state's court.

Engineering societies resemble craftsmen's guilds. To become a junior engineer, one must first be educated in a university with an accredited engineering program. However, having the knowledge acquired at such an institution is not enough. All engineers subscribe to the adage: if you build it, it must work. Many 'apprentice' engineers work for a corporation where they become part of a guild-like engineering community. Apprentice engineers learn how to design, validate and build things that work. Unlike religious and social communities, acceptance and relationships are not just built on creeds or behavioral norms; they require skills.

Scientific communities resemble the Islamic warraqeen in that they submit their work to other scientists for review. In his book, *The Structure of Scientific Revolutions*, Thomas S. Kuhn points to the importance of acquired skills: "Tacit knowledge results from learning by doing science and has the following characteristics:

- It has been transmitted through education.
- It has, by trial, been found more effective than its historical competitors in a group's current environment.
- It is subject to change through further education and through discovery of misfits within the environment.
- It misses one characteristic of knowledge: there is no direct access to what is known, no rules or generalizations with which to express knowledge. (e.g., electrons are not seen, but rather their tracks).
- It requires interpretation which begins where perception ends; what perception leaves for interpretation to complete depends drastically on the nature and amount of experience and training."

A scientific theory is usually felt to be better than its predecessors if it is a better instrument for discovering and solving puzzles, and representation of what nature is really like. Theories are built on paradigms composed of:

Symbolic generalizations, deployed without dissent, can be readily cast in a logical or symbolic form (e.g., $f=ma$). The power of a science generally increases with the number of symbolic generalizations.

Beliefs supply preferred or permissible models of shared analogies or metaphors (e.g. heat is the kinetic energy of the constituent parts of bodies).

Values judge whole theories or predictions which:
- permit puzzle-formulation and solution,

- are simple, self-consistent, plausible and compatible with other theories,
- assure accuracy: quantitative predictions are preferable to qualitative ones,
- identify margins of permissible error that should be consistently satisfied.

Examples provide concrete problem-solutions which:
- are found in periodical literature,
- are encountered by students in laboratories, on examinations, and in texts,
- allow students to see a variety of situations as alike,
- allow scientists to solve puzzles by modeling them on previous puzzle-solutions,
- allow scientists to see the same things when confronted with the same stimuli (e.g. physicists recognize tracks of alpha particles and electrons),
- provide empirical content for the laws and theories a student has previously learned,
- provide symbolic generalizations (e.g. in Newton's 2nd Law of Motion is a law-sketch or law-schema; f=ma in free fall becomes $mg=d^2s/d^2t$)."

Scientific communities require belief in its paradigms or dogmas. They have success in predicting physical events, much less in predicting biological events and no success in predicting human and societal events. Except in Economics, the idea of value is not part of science. Religion explains the value of human experience for the individual and how he is to behave in community.

Most major religions have two communities: the faithful and its leaders or interpreters who may be called priest, pastor, bishop, rabbi, monk, imam, etc. Depending on the religion, the faithful may have more or less to say about interpreting God's word from sacred texts.

Community and grace are central to Christianity. At its best, it is a religion of grace, comfort, forgiveness and a powerful force for good. At its worst, it invoked the murderous behavior of the crusades. In Orthodox Christianity, belief is critical for redemption, not reason. Christian and Jewish communities were not always simply structured on orthodox belief, but sought God's truth.

Princeton Professor of Religion Elaine Pagels points out that Orthodox Christianity was not the only view for three centuries after Jesus death. Gnostic Christians accepted Christ's teaching, but not His divinity. They believed that one comes to God through knowledge of self and the world and, through meditation, knowledge of God. It allowed for adaptation as their knowledge expanded.

Matthew and Luke follow Mark in describing Jesus as a future king (messiah, Son of God); a mortal invested with divine power (son of man). John's gospel contradicts their testimony: Jesus is God himself revealed in human form.

Pagels compares the Gospel of John with the writings of Thomas. John writes "so that you may believe, and believing, may have life in Jesus' name." Thomas encourages the hearer to seek to know God as did Jesus through one's own, divinely given capacity, since all are created in the image of God.

Irenaeus (Bishop of Lyons, 2nd century) argued militantly for four gospels: Mathew, Mark, Luke and John. Irenaeus resolved to hack down… "apocryphal and illegitimate" writings (of Gnosticis) leaving only four pillars standing. He claimed that these writings were trustworthy, because Jesus' disciples Matthew and John actually witnessed the events they related. Few New Testament scholars today agree with Irenaeus (Mark: c. 68 to 70 CE, Matthew and Luke: c. 80 to 90 CE, and John: c. 90 to 100 CE).

In the 3rd century, Constantine believed he had found in Christ an all-powerful divine patron and the promise of eternal life; he legislated the moral values he found in biblical sources, built upon divine justice. He enhanced the authority of the bishops identified as catholic and established their consensus expressed through the creed.

Jewish, Christian and Muslim doctrines of God separate the divine from what is human and assume that divine revelation is opposed to human perception. They rule out spiritual truth that may come from human intuition, reflection, or creative imagination (like science).

Astrologers, alchemists, masons and witch doctors have evolved into modern day scientists. Past religious leaders were the philosophers of their day and tried to explain the ethics of human behavior, man's relationship to God and God's work in our environment. Christian scholars continuously expand their "rational" defense of its 1500 year-old static dogma. Whereas, science communities continuously rebuild their symbolic generalization, beliefs, value systems and examples.

Consider a new theological paradigm derived from Kuhn, the spirit of Gnostic Christianity, Hindu spiritual transcendence of self, modern guild structures and the evolution of the aesthetics of worship to provide the foundations for Pagel's hope that people will validate themselves in religious experience. Let us paraphrase Kuhn's definition to a theological paradigm for a religion.

A religion's success is measured by its number of adherents whereas a theology is usually felt to be better than its predecessors if it is a better instrument for discovering and explaining what the nature of man to God is really like. Whitehead's primordial and consequent nature of God is such an instrument since it explains many disciplines: quantum physics, relativity, cosmology, evolution, sociology, psychology, philosophy and religion.

Theologies are built on paradigms composed of:

Symbolic generalizations deployed with little dissent by its members that can be cast in mystical form, but do not contradict reason or what is known from history or science. A theology's power increases with its ability to explain what is known and imagine what is unknown.

Beliefs that supply the group with preferred or permissible models of shared analogies (e.g. theistic evolution).

Values that judge the whole theology or concern human behavior which:
- permit codes of conduct for ethical behavior,
- are simple, self-consistent, plausible and compatible with a priori moral knowledge,
- assure ethical behavior: specific responses are preferable to qualitative ones,
- identify margins of permissible error that are acceptable for human behavior within the community.

Examples that provide problem-resolutions (parables, analects, stories, myths) which:
- are found in scientific as well as religious texts,
- are encountered by seekers/believers in other texts,
- allow seekers/believers to see a variety of situations as like each other,
- allow seekers/believers to resolve ethical dilemmas,
- allow seekers/believers to see the same resolution when confronted with the same issues (e.g. surprise pregnancy, end-of-life disease, stem cell research),
- provide empirical content for the theology a student has previously learned.

Before describing a theological community consider what the Apostle's Creed of the Christian church calls the "communion of saints." Saints in the New Testament refer to baptized believers in a local congregation, such as the saints at Corinth, or Ephesus, or Colossae. The New

Testament recognizes "a great cloud of witnesses" who surround us as we walk the path of faith.

Each Christian tradition can offer its own special "saints" to the entire body of the church. In this way one can respond to the biblical call to the church's doctrines. The communion of saints acknowledges that all Christians, past and present, participate in the religious community. It evolved from Judaism's strong sense of community.

For science and engineering, in the practical application of written narrative, mathematics, prototypes, theories, and well-defined experiments there is a 'communion of saints' where Newton, Einstein and others continue to contribute.

Theological communities must answer the questions (not to be confused with church or religious community):

- How does one elect and how is one elected to membership in this community?
- What is the process and what are the stages of socialization to the group?
- What does the group collectively see as its goals?
- What deviations, individual or collectively, will it tolerate?
- How does it control the impermissible aberration?

Many of today's religious communities barricade themselves behind the details of errant dogmas. A community of respected thinkers (scientists, philosophers, theologians) must convene to develop a framework for and components of an ecumenical theology. Its initial purpose would be to clarify and identify the harmonious and cancerous components of religious dogma.

Once a community validates revelations, the faithful may take on a project of expanding their community. This is accomplished using elements of a religion that contain one or more of these elements:

Symbolic Generalizations	
Ritual	habitual performance of actions that reinforce religious elements: recitals, songs, readings, lectures...
Artifacts	objects, statues, architecture, structures and icons to appeal to the emotional acceptance of religious elements.
Beliefs	
Spirituality	belief in a Being greater than both self and humanity.
Mythology	imagined Being (God).
Mysticism	communication with that Being *(prayer, meditation)*.
Incentives	prosperity, serenity, transcendence, afterlife, hell, heaven.
Values	
Ethics	behavior, law and societal norms of value and morality.
Sacred texts	religious elements synthesized into a general system.
Examples	
Parables	metaphors that explain abstract ideas by analogy.
Proverbs	pieces of advice designed to impart wisdom.
Analects	instructions that can be followed without interpretation.
Myths	stories that imagine the Ultimate Being.
History	records and stories about how and why religion evolved.
Cosmology	answers to fundamental questions about life and the nature of the universe.
Music	reinforcement of examples, values and beliefs.

In recent years organizations have recognized that their most valuable asset is employee skills, experience, education and knowledge. Maintaining and improving this asset is critical to the capabilities and profitability of organizations. The formalization of this process is called knowledge management (KM). KM is seen as an enabler of organizational learning. A key element of KM is CoPs.

A way to expand the faithful's community is to embed itself as a Community of Practice (CoP) in society. CoPs are groups of people who share information, insight, experience, and tools about an area of common interest. They facilitate internal and external communication to determine best practices and transfer organizational knowledge within and across divisions and disciplines. CoPs create an environment for socialization in which knowledge is created, validated and shared.

Religious CoP Charter

Community of Practice: Embed community in society.

Charter Statement: The community will interact with society to influence the ethical behavior of those not in the community and encourage them to become part of it.

CoP Support: Those individuals that support community planning and related societal activity.

CoP Stakeholders: Ministry and community leadership.

CoP Members: Community of the faithful members that regularly attend and contribute to the CoP's meetings.

Measurable Objectives:
- Instill beliefs in the community of the faithful: increase the faithful that understand its theology.
- Continuously improve and use an ideology for societal analysis: size of membership.
- Illustrate and promote the ethical life for all in society: decrease in the total number of felonies.

CoP Activities to Support Objectives:[12]

After a theological community develops or refines its beliefs, its members are responsible for ministry (the act of serving) to a community of the faithful. All ministry occurs within the community; moreover, its perspective broadens to incorporate a society in which it is embedded.

Communities of faith relate to a wider society in diverse ways, which may or may not share the community's faith and religious assumptions. Conversely, a religious community may or may not share many of the political social values common in the wider society. The question to ask is: in what ways does the religious community relate to the wider social world of which it is a part and

---

[12] J. E. Bush Jr., Practical Theology in Church and Society 2012

how might the two be related at any given time and in any given place? Placing these two directions of inquiry together, the figure conceptualizes how a community of faith organizes itself to relate to the rest of society and how religious beliefs are shaped by and give shape to ideas and values circulating in society.

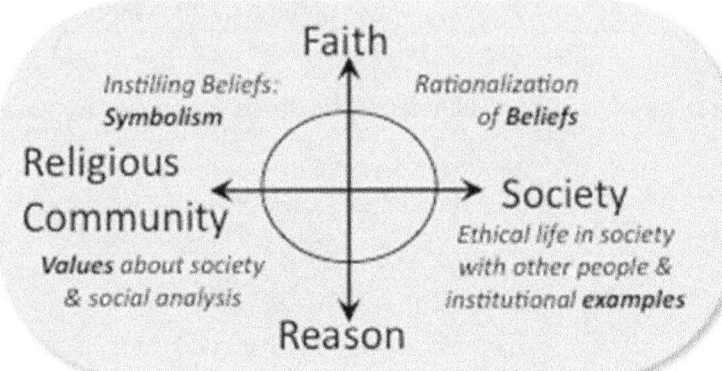

Each quadrant in the figure represents a different aspect of this relationship. The lower right quadrant represents *life together in society* - who the faithful are as social beings economically and politically. This quadrant represents the myriad of social forces that shape the community. The community starts to live together in the same social conditions that unite and divide the community and society.

Values in the lower left quadrant are how the faithful think about life together in society. In reasoning about social practice, the faithful engage in social and cultural analysis. But values and examples compete for people's attention, offering stories, models, explanations or justifications of the social forces at work in the society.

This analysis of society digs deeper to discover political power, economic forces, institutional arrangements, cultural patterns, community problems and assets, and social struggles and resources. For a particular community, how do beliefs and assumptions reflect other important patterns of thought and values in the shared social world? How can theological reflection be informed more deliberately by social analysis? How might religious thinking enrich the larger shared world of meaning?

These questions about *instilling beliefs* move into the upper, left quadrant: theological understanding, religious reflection, and faith assumptions about church, society, ministry, God, etc. It brings attention to the community's religious elements. But, at the same time, it relates religious practice to the wider society and to the community's own practices in ministry.

Rationalization of beliefs in the upper right quadrant represents the particular practices of the community. The practices rationalize the concrete reality of the faithfuls' life together in the world. It includes the institutional and social form of the community, the actual practices of ministry by the faithful in community and society, and the many ways in which to serve neighbors in society. The practice includes ways in which the community might want to separate from or cooperate with society.

Community Activities to Meet the Challenges

This method develops ministerial practice that is more socially relevant or theologically informed and looks for bridges between each of these quadrants as well. The method, frequently, begins with the social situation itself in the lower right quadrant. This point of entry allows one to take the perspective of community in society.

1] How do the faithful enter a new society, or how do they take their own social context or social location seriously? These questions situate the faithful within a social context.

2] In the lower left quadrant, analyzing forces at work in a social context: ideological, economic, political, cultural, or familial; and patterns of oppression such as classism, racism, slavery, sexism, colonization, immigration or war.

3] Explicit theological reflection connects between the faithful (its sources, symbols, traditions and breadth) to the social situation it has been analyzing. How do social analysis insights correlate to the theological perspective?

4] Finally, how does exploring theology in light of social analysis move one to think about ministry? How might one need to change, persevere, sever or build alliances? How might one need to expand social outreach? How might the community, ministry, or society be involved? How does ministry affect or transform the lives?

Author's Reflections

Atheisms success in eroding the expression of religion in the public square continues to be a detriment to society. Although it does not necessarily follow that atheism produces abhorrent destructive societal infrastructures, we have witnessed the brutality of communism where atheism becomes the state religion.

It is time for a religious ecumenical community to confront this destructive trend. Constantine in the 3rd century and the USA's founders understood the impossibility of administering a dictatorship or a democracy without a preponderance of religious belief among its citizenry.

Leaders (ministers, priests, imams, bishops, etc.) of many religious communities today claim authority over sacred texts and assume the responsibility for their interpretation. They dictate how the faithful participate in society. Many religious leaders fail to seek God's new revelations in science and art that revise perceptions of our world and universe; and fail their faithful asking for too much belief absent reason.

Religious communities share symbolic generalizations and beliefs, but values and examples tell its community how to behave in society. Sacred texts rationalize beliefs and rituals into a comprehensive system of ethics to guide behaviors and societal norms, values and morality. The examples – parables, proverbs, analects, myths and cosmology - attempt to answer fundamental questions about life and behavior. History, parables and cosmology need continuous updating based on new scientific and religious revelation.

Laws for responsible citizenship and penalties for failure may not agree with some elements of religious culture. Ultimately, every individual is responsible for his or her own behavior and must question religious interpretation and authority especially when it preaches violence.

Sacred texts are a part of all world religions. They must be read and understood in the context of the times in which they were written. Regrettably, people interpret them with a personal agenda and use sacred texts out of context to support unsubstantiated beliefs. Others treat these texts as God's inerrant word rather than man's interpretation of God's truth; hence, the vast different views of God in major religions. Unquestioned belief in anything to ensure the faithful a superior place in eternity has been and is being used to perpetrate many atrocities.

Both an inter-religious ecumenical community and local religious Communities of Practice are necessary for religions to take their effective and rightful place in society. Both communities must constantly monitor and question the behavior of its members and leadership in their interpretation and implementation of the elements of their religions. The next chapter will explore elements of world religions.

Religion

Religion infused society with values derived from revelation.	
• Images of God	• What are the elements of a religion?
• Worship	• How did religions evolve?
• Man's religions	• What are the major world religions?
• Belief	• What comprises a rational religion?
• Soul	• What role do sacred texts play in religion?
Religious validation has tended toward extremes: all new ideas are summarily rejected if they contradict sacred texts; conversely, new age ideas from a charismatic leader are accepted without question. Religious validation is key to evolving systems of worship and values that respond to a rapidly changing world order (population, technological advances, ecological degeneration, etc.). Religion imagines God as that force which created and sustains the universe and speculates about questions like: How did the universe get here? What is the meaning of life? What happens after death? What does God want? Because religions are based on man's fallible, divinely inspired, imagined view of God, sacred texts represent extraordinary diversity in beliefs.	

Although some persons' aim most at intellectual purity and simplification, for others RICHNESS is the supreme imaginative requirement. When one's mind is strongly of this type, an individual religion will hardly serve the purpose. The inner need is rather of something institutional and complex, majestic in the hierarchic interrelatedness of its parts, with authority descending from stage to stage, and at every stage objects for adjectives of mystery and splendor, derived in the last resort from the Godhead who is the fountain and culmination of the system. - William James

How is a validated revelation brought to the masses? "God's Revelation" as interpreted and validated by a theological community is communicated to the masses through religious formation and use. This chapter reviews the history of communicating "Revelation" to the masses through organized religions.

The great religions of the world are built upon the word of their founding prophets and mystics. Their many forms of worship are meant to educate and assure the followers through their reason and emotion. Worship appeals to the faithfuls' aesthetic, emotional and intellectual senses with authority, mystery and splendor. Religions imagine God in many and varied forms. Several imaginings include:

- Angry Judge
- Benevolent Father
- Savior Son
- Unnamed / unspoken force / spirit
- Flow of Nature
- Universe Creator / Maintainer

From the previous chapter, religion contains one or more of these elements:

Symbolism	Beliefs	Values	Examples	
Ritual	Spirituality	Ethics	Parables	Myths
Artifacts	Mythology	Sacred texts	Proverbs	History
	Mysticism		Analects	Music
	Incentives		Cosmology	

Major religions evolved because their faithful took on a project of expanding their religious community. To understand how this is accomplished, the chapter considers how religions evolve, what are the major religions and what comprises a rational religion. It also considers what roles sacred texts must play in the evolution of religion to support the ethical issues of community and society.

The *Tao Te Ching* describes God as "The Way that can be spoken of is not the constant way; the name that can be named is not the constant name. The nameless was the beginning of the heaven and the earth." With no idea of what God is, a discussion of religion continues.

In his book, *Discovering God*, Sociologist Dr. Rodney Stark asserts that if certain rare individuals have the capacity for revelation to create profound new religious truths, from real communications, from the supernatural or from an unusual creativity and if religions have much in common, then it can be argued that is to be expected because each is based on revelations from one true God.

In his book, *The High Gods in North America*, Jesuit priest anthropologist Wilhelm Schmidt in 1933 presented his theory of primitive monotheism: primitive religion among tribal peoples began with an essentially monotheistic concept of a High God — usually a sky god — who was a benevolent creator. Schmidt theorized that human beings created a God who was the First Cause of all things and Ruler of Heaven and Earth before men and women began to worship a number of gods.

The 'fall' from primitive monotheism coincides with the rise of civilizations – societies having cities and productive agriculture. "Starting in 3100 BCE, separated by geography, Sumer, Egypt and Greece and later Maya and Inca maintained elaborate state religious polytheisms controlled by priests and sustained a similar array of gods. Housed in great temples, these religions were fully funded by land grants or subsidies; priests were in service to despotic rulers, some of whom presumed to be God."

A shift back to monotheism occurred in the 6th century BCE with the Buddha, Confucius, Lao Tzu (Taoism), Zoroaster, Mahavira (Jainism), the authors of Hindu Upanishads, Pythagoras (founder of Orphism in Greece), and Israelite prophets: Jeremiah, Ezekiel, and the second Isaiah. These religions did not emerge initially as monotheistic, but all required moral behavior.

Hinduism arose in India during the fifteenth century BCE. It changed profoundly in the sixth century BCE. Scriptures of the Upanishads introduced the doctrine of the transmigration of souls – upon each death one is reborn and life's unification is determined by Karma. Immoral behavior is punished and good behavior is rewarded. Similar ideas arose at this same time in Zoroastrianism and Judaism. The central feature of the New Hinduism was the self or "atman," and how it linked to Brahman, the unified spiritual force. Like the Christian soul, the atman is eternal while the body is mortal.

Hinduism's Upanishads culminates in a new scripture: Vedanta. It's monotheism stresses: Brahman beyond comprehension and Isvara within Brahman that is the creator, maintainer and preserver of the universe.

Jain doctrines were similar to Upanishad scripture, except Jains do not believe in a Creator God, since their universe has no beginning or end, but passes through an infinite number of cosmic cycles.

Siddhartha Gautama founded Buddhism in the sixth century BCE. Jainism and Buddhism both deemphasized God, but Buddhism achieves enlightenment through meditation while Jainism favored extreme self-mortification. Buddhism expanded Upanishad themes. [13]

By the 11th century CE, Hinduism's Vedanta had evolved to include elements of Jainism and Buddhism. Vedanta's most fundamental teaching is that all that exists is divine. Every human being is innately divine whose ultimate goal of life is to manifest this inherent divinity. Divinity is equally present, but not equally manifest everywhere. In humans, divinity is most manifest in a spiritually

---

[13] John Cooper: *Panentheism: The Other God of Philosophers*
Hinduism: Sarvepalli Radhakrishnan (d1975) developed a panentheism that seeks common ground with Western philosophy; and non-Hindu religions building on the works of Hindu philosophers: Sankara (d820) who taught that God is absolutely one and that all distinctions and differences are merely temporary illusions; and Hindu Ramanuja (d1137) held that the world is the body of the Brahman (God), that individual souls are real, and that souls do not disappear into God.

Zen Buddhism: Masao Abe, a Japanese Zen master and Alan Watts, an Anglican priest who became a guru for Zen Buddhism in the 1960s, found common ground in Western panentheism. Zen Buddhism is a convergence of Buddhism and Taoism and Confucianism in China. Unlike the Western tradition, Zen does not favor the triumph of being over nonbeing, but accepts both in perfect balance. By design, Watt's Zen Buddhism is panentheistic.

illumined soul. From John 8:12, Jesus said "I am the light of the world: he that follows me shall not walk in darkness, but shall have the light of life."

For thousands of years, China's religions stressed ancestor worship and performed elaborate rituals to them. All classes believed in divination and called upon various nature gods, according to their needs. Taoism and Confucianism entered in the sixth century BCE.

Lao-Tzu wrote the *Tao Te Ching* in 81 brief paradoxical chapters in keeping with the warning: "He who knows does not speak. He who speaks does not know." The Tao can be translated as "The Way." It is beyond being and inexpressible: "The Tao that can be told is not the eternal Tao." Tao is the basis of all being, the uncaused cause.

Confucius's teachings "devise a set of moral and political beliefs which, if widely and consistently acted on in human society, would produce the greatest good: ...a stable social and political order in which human beings could flourish." Unlike Machiavelli's cynically amoral *The Prince*, Confucius's Analects urge exemplary morality and propriety whereas Jesus' parables urge love, morality and forgiveness.

Buddhism brought to China from India between 100 BCE and the first century CE was transformed by contact with Zoroastrianism as it passed through Persia. Indian Buddhism Nirvana is a state of nonbeing releasing the individual from the endless cycle of death and rebirth. Chinese Buddhism Nirvana is a heavenly place where "people get whatever they wish for." The state dismantled both Buddhism and Taoism in the eighth century CE, but they are now part of Chinese folk religion combining Taoism, Confucianism and Buddhism.

Judaism is the religion of the book; *Tanakh* (the Christian *Old Testament*) plays the authoritative role in religious life. "Deuteronomists" in the 6th century BCE converted all of Israel to monotheism, the exclusive worship of Yahweh, revising many parts of the Bible accordingly. Exiled in 597 BCE in Babylon, Ezekiel and the second Isaiah shaped the religious life of this community. Ezekiel's prophecies contain Zoroastrian influences: notions of heaven and hell and Satan who causes calamities and forces humans with "free will" to choose between good and evil.

Five centuries later, three Judaic religious sects emerged: Sadducees, Pharisees, and Essenes. Sadducees, temple priests, rejected "fate" in human affairs, postulating that humans had the power and responsibility to determine their own actions; they denied immortality and resurrection and taught that God's rewards are gained only in this life. A Sadducees high priest judged Jesus guilty of blasphemy.

Pharisees believed in the immortal soul, resurrection of the good, and condemnation of the wicked in eternal torment; "the good" were those who obeyed written and oral law. Essenes were a reclusive, pious ascetic sect that condemned "pleasures of evil," rejected marriage, and embraced abstinence. Jesus teaching and behavior suggests that he combined elements of the Pharisees and Essences while speaking out against the Sadducees.

Jesus' unique message was that all could be saved which entailed an attractive afterlife. It was not reserved for the elite, but was for anyone who believed in Jesus' teachings and was baptized. Religious life was far more attractive to females since they were regarded as equal to males in the eyes of God.

Christianity's revolutionary aspect lay in moral imperatives such as: "love one's neighbor as oneself; do unto others as you would have them do unto you; it is more blessed to give than to receive; when you did it to the least of my brethren, you did it unto me." Its response to long-standing misery of life in antiquity offered people the compelling "material" reasons to convert.

Early Christianity was both a religion of the book and a theological religion. It does not rely merely on stories, sermons, sayings of Jesus and stories of miracles. It stood on general principles of morality, reality and the nature of God. It was deeply rooted in formal reasoning, Judaic law and Greek logic. [14]

In 312 BCE, Constantine embraced and converted to Christianity. Christians were a substantial majority in Rome and other cities. He was not responsible for the triumph of Christianity; rather Christianity provided him with substantial and well-organized urban support for his struggle to gain the throne. Pagan conversions to Christianity were greatly facilitated by extensive cultural

---

[14] John Cooper: *Panentheism: The Other God of Philosophers*

Christianity: Pierre Teilhard de Chardin, S.J. (d1955) synthesizes his devotion to Christ, his understanding of Roman Catholic theology, and his evolutionary cosmology into a grand scientific-metaphysical-mystical vision. Many regard Teilhard as a spiritual genius who intelligently and effectively expresses the essence of Christianity for the contemporary world. Christians and non-Christians work from his panentheistic view of God, the cosmos, and the future of the global community.

Judaism: a Hasidic Jew, Martin Buber (d1965) in his book, *I and Thou*, outlined a relational-dialogical ontology that includes nature and humans in God. Buber presents a panentheism in which God transcends and includes humanity and the world by addressing us as 'You,' an eternal word that constitutes the very being of God and creatures in a relationship where humans can respond with 'You.' He desires a living relationship with the personal-covenantal God of the Hebrew Scriptures.

Islam: Mohammed Iqbal (d1938) was politically and intellectually active in developing a modern Islamic worldview. His roots are in the Sufi tradition, but his view is panentheistic. He portrays God and the world as infinite soul and finite body. The world and everything in it contribute to the becoming of God, the realization of the infinite creative possibilities of His being.

continuity between classical paganism and elements of the Christ story of blood sacrifice and the virgin birth.

Group	Religion	Adherents	Origin	Main regions	View of God
Abrahamic 3.4 billion	Christianity	2. billion	1st c.	Worldwide except parts of Africa & Asia.	Trinity: Father, Son, Holy Spirit
	Islam	1.4 billion	7th c.	Worldwide except parts of the Americas.	Allah: One Father God
	Judaism	14 million	1500 BCE	Israel, USA and Europe	Jehovah: One Father God
Indian 1.4 billion	Hinduism	0.9 billion	1500 BCE	India, Fiji, Guyana and Mauritius	Brahman, Vishnu, Shiva
	Buddhism	0.4 billion	600 BCE	India, East Asia, Indochina, Russia.	Nirvana: beyond comprehension
Chinese folk religion 0.4 billion	Taoism	0.4 billion or more combined	600 BCE	China	Tao: Way of the Universe
	Confucianism				Unnamed
	Buddhism				Nirvana: beyond definition
Ethnic/ tribal 0.4 billion	Primal indigenous	0.3 billion	?4000 BCE	India, Asia	Ancestors and Pagan
	African traditional	0.1 billion	?4000 BCE	Africa, Americas	Ancestors and Pagan
Irreligious		1.0 billion			Secular Humanism

A demographic distribution of modern major religions is shown in the table above (ref: www.adherents.com).

Muhammad, who died in 632 CE, founded Islam. By 714 the Muslim Empire stretched past Tangier into most of Spain. Islam is easy to learn and to follow which greatly facilitated its spread. Like Judaism, it is a religion of orthopraxy, or correct practice, in contrast with Christianity's orthodoxy, or correct beliefs or doctrine.

Islam is based on Five Pillars plus one:
- Confession: "I bear witness that "There is no God but Allah; Muhammad is the messenger of God."
- Worship: praying five times a day facing Mecca.

- Giving alms to all those in need at year's end for Allah.
- Fasting from dawn to dusk during Ramadan.
- Pilgrimage to Mecca.
- Shari'a law: a full legal system for a Muslim state.

In his book, *The Making of Religion*, Whitehead explains that religions exhibit four characteristics: ritual (organized procedure), emotion, belief, and rationalization (adjustment of beliefs into a system). In primitive religions, belief and rationalization are negligible. When belief and rationalization are well established, solitariness constitutes great religious conceptions: e.g. meditations of the Buddha, Christ and Mohammed.

Rational religion is religion whose beliefs and rituals have been reorganized to make them the central element in a coherent ordering of life. It is coherent both in respect to the elucidation of thought, and to the direction of conduct toward a unified purpose commanding ethical approval. Most modern religions have evolved to rational religions.

When Constantine adopted Christianity, it had the widest possible view of social structure for a communal religion. The Roman Empire's religion, based on Christianity, reinforced the power of the church's Bishops and imbedded it in law, first by Constantine, then by Justinian in 530 CE extended into the 17th century. A university law school may have constructed it today, impressed by the notion that imprisonment as a deterrent to crime pales in comparison to eternal damnation.

In 300 CE, there were many rival religions; but only Buddhism and Christianity combined the clarity of idea, generality of thought, moral respectability, and width of extension over the world. Later, Islam challenged their position. Their authority is now challenged by science.

Science attempts to formulate in precise terms the truths disclosed in the sense perception of mankind. In the same way, religion attempts to formulate in precise terms the truths disclosed in religious experience. Sacred texts should be founded on:

1] the value of an individual for himself;

2] the value of individuals in community for each other;

3] the value of the objective world which is a community derived from the interrelations of its component communities, and their individuals.

Modern secular society has lost God and is seeking him. Simplicity of religious truth is a favorite axiom of some theologians. It bases religion on a few ideas, which are most effective in producing pleasing emotions and agreeable conduct. Physics and genetics do not disclose a simple world. If our trust is in the ultimate power of reason for the discernment of truth, we have no right to impose simple conditions in view of the problem of evil.

Although God is beyond human understanding, Swami Krishananda's says that the *Mandukya Upanishad* explains human nature and the attainment of oneness with the Absolute (God, Brahman). Atman (soul) is immortal essence in all living organisms. The purpose of each individual's Atman is to engage in a process of transcendence to realize oneness with Brahman.

The jiva (immortal consciousness in all living organisms) tries erroneously to release tensions by fulfilling desires; they cannot be fulfilled by contact with objects, because contact excites a further desire for a repetition of contact, which, in turn, excites an additional desire, in an endless cycle. Desires arise from ignorance of the structure of things. All things perceived in the waking state are false, Reality and Truth are realized when the jiva's Atman ascends in four stages to be one with the Absolute.

Stage 1) Waking consciousness studies, observes, experiments and deals with other objects and persons.

The self is analyzed from a foundation of sense-perception and mental cognition. To humans, whatever is presented in the waking state is considered to be real. Jiva's delusion and undoing is a result of evaluating, judging, wanting or not wanting the world in some way.

Stage 2) In the subtler function of dream sleep, objects of dream are our own mental creation. Witnessing both states in this stage is independent of waking and dream. Western psychologists conclude dreams are due to personality complexes: Freud attributes them to sex, Adler to feelings of inferiority and Jung to an urge for growth and harmony. Suppressed desires become psychological complexes, which may become diseases. Dreams approximate reality; but, the waking ego subsides and does not oppose healing forces. In dream, Isvara's higher powers may operate.

Stage 3) Isvara (Universal Consciousness) is deep sleep where all perceptions and cognitions converge and comingle into a single mode of mind instead of many cognitive psychoses. Isvara is the manifestation of Brahman that human consciousness can comprehend. Deep sleep is filled with bliss, delight, and satisfaction. Through Isvara, the jiva has the capacity to understand, rationalize and judge situations, but Isvara's law judges as well. If our egoism violates His law, the world will recoil upon us.

Stage 4) Brahman (Transcendent Presence), the fourth state of the Atman, embraces the first three states of manifested consciousness, and also Brahman: creator, preserver and destroyer; omnipresent, omniscient and omnipotent; 'That which is' said Saint Augustine. This is the Atman where there is neither waking, nor dreaming, nor sleep. Cessation of the ego is the revelation of the Atman. God and the true measure of Reality are revealed.

There is a quality of life, which lies always beyond the mere fact of life. Religion is the direct apprehension that, beyond happiness and pleasure, there remains the function of what is true, that it contributes its quality as an immortal fact to the knowledge that informs the world.

To illustrate a religion's view of God consider the Hindu's interpretation of the Lord's Prayer. In *The Sermon on the Mount according to Vedanta*, Swami Prabhavananda tells us: Every great teacher, no matter whether he is a Divine Incarnation, prophet or illumined soul, has two sets of teachings one for the multitude and the other for his disciples. The teacher of religion gives the inner truth of religion only to his intimate disciples. A truly illumined teacher can transmit to us the energy and impulse, which makes us unfold within ourselves. The disciple must make himself ready and prepared.

In the Sermon on the Mount, Christ talks to His disciples who could grasp what He was teaching them not merely through His words, but also because He was personally transmitting something to them for which their hearts were prepared. The Lord's Prayer begins: *"Our Father which art in Heaven, Hallowed be Thy name."*

According to Prabhavananda, all world religions teach us that heaven and earth is within. In the three states of consciousness waking, dreaming, and deep sleep we have a sense of identity and continuity; it's like a river, always changing, ever flowing. Heaven is that which abides forever. We must first recognize Heaven within ourselves; then we find it everywhere.

Christ tells us that our Father is in Heaven. In every age, people ask for proof of God's existence. Proof by reason is futile because trying to prove the existence of an idea of God does not guarantee that the idea and the Reality of God would correspond. Throughout history, there have been many great teachers and saints and Illumined Souls, who have told us: "God exists. I have seen Him." If we believe that these men were not deluded, then a conviction begins to grow in our own hearts. This book does not reveal God, but provides a concept of God that provides a framework to underlie science and religion in the form of Whitehead's panentheism.

Prabhavananda considers the words "Our Father." Christ is teaching us how to think of God when we pray to Him. We are to think of Him as our very own Father, whom we regard with awe and reverence but chiefly with trust, confidence and love. We are safe under His protection.

Sri Ramakrishna said that God is personal, impersonal and beyond. "Beyond" being is another absolute reality of God, which no mind can grasp. However, the religion, which Christ taught is what Hindus call Bhakti Yoga, the path of loving devotion. To a devotee of this path, God, as an impersonal being, is too much of an abstraction to be loved. In order to be able to pray and meditate, he must enter into a relationship with God (Isvara) as a person.

"*Hallowed be Thy Name.*" In almost every religion we find that emphasis is laid upon the Name, the Logos, the Word. In Vedanta, God's name is called up in the Mantra. The teacher gives a Mantra to every devotee at the ceremony of his initiation. By repeating this Mantra, hallowing His name, over and over, one lets God take possession of his conscious mind.

"*Thy Kingdom Come. Thy Will be done in earth, as it is in Heaven.*" When a Hindu performs ritualistic worship, the first prayer he says is: "As with eyes wide open, a man sees the sky above him, so the Seer sees God continually, the Supreme Truth, the All-Pervading Existence." God's Kingdom has come already: it has always been with us. So Christ teaches us to pray that we may transcend space and time and may know God's Kingdom while on earth.

"*Give us this day our daily bread.*" This is the bread of Grace. We pray that this Grace may be revealed to us. We are asking to have it "this day at this very moment." Your own struggles will never make you pure, but we must also know that God's Grace may descend at any moment.

"And forgive us our debts, as we forgive our debtors." A Hindu who reads this passage will at once understand the word "debt" to mean the debt of Karma. When we realize that everything, good or bad that comes to us has been previously earned by ourselves alone, then we know that we must not hold anybody else responsible for anything that we suffer in our lives. From that point, we can go on to say: "If it is by my own doing that I am what I am, then I can also become what I wish to be."

"And lead us not into temptation, but deliver us from evil." How is it possible for God to tempt anybody? The Gita replies that this whole universe is one gigantic temptation, and that this universe is Maya, the Divine Illusion. In the story of the Garden of Eden, Adam is warned not to eat the fruit of the Tree. When he eats it, his Atman identifies with his ego: he recognizes good and evil and the universe of Maya. Ego is the root-cause of ignorance. To escape from Maya we surrender the ego completely to God: *"the kingdom, and the power, and the glory, for ever."*

Religions insist that our knowledge of ordered relationships, especially in aesthetic valuations, stretches far beyond anything that has been expressed systematically. This is the point that science forgets.

Dogma in sacred texts adequately explains a limited number of abstract concepts. Religions fail by insisting that dogma is absolute truth. Religious inspiration is found in expression of the finest types of religious lives. Though some supreme expressions lie in the past, sources of religious belief are always growing. They are not formula, but elicit in us a response that extends beyond dogma.

Panentheism offers a compelling concept of God that has been espoused by a prominent philosopher in each major religion. Unfortunately, their religions failed to adopt it in their "understanding" (or more aptly to minimize their misunderstanding) of God. Whitehead's panentheistic view of God is used in the next chapters to synthesize a framework that embraces science and could revise as well as reaffirm the dogma of major religions.

According to Hartshorne's process theology based on Whitehead's panentheism, without God, the world would be nothing more than a static, unchanging existence radically different from the actual world of experience. An eternal and temporal God provides possibilities to change and develop the world.

God's eternal objects (akin to Plato's Forms) provide an actual source of possibilities. However, if God were only external, His possibilities would be unrelated to the actual world, as it presently exists. For God to be related to the world, God must be present in the world and the world must influence God. Process theology avoids collapsing the world into God or God into the world (pantheism) by distinguishing an eternal God and a material world.

Although God presents possibilities to events in the world, because the world's occupants have free will, each event "decides" how it will realize its possibilities. Since God does not determine the response of each event to the possibilities that He presents, any event may reject God's purpose of good to possibly produce evil. Humanity (our collective jivas) is the source of evil not God.

Author's Reflections

The failure of religions is demonstrated by the rise of secular humanism and atheism in the US and Europe's public square. Seminaries of Christianity persist in reinforcing 1500 year-old dogma and the absolute truth of sacred text at the expense of theological scholarship. Consequently, atheists argue that religion should not guide societal ethics. As a secular society replaces religious mores, it must increasingly legislate ethical behavior.

Speaking on the importance of religion Vivekananda says: "the mainspring of the strength of every race lies in its spirituality, and the death of that race begins the day that spirituality wanes and materialism gains ground... when we come to the real spiritual, universal concept, then and then alone will religion... live in our every movement, penetrate every pore of our society and be infinitely more a power for good than it has ever been before."

In his book of fiction, *Life of Pi*, Yann Martel describes a scene where a priest, imam, and pandit with whom Pi had been practicing religions approach Pi and his parents. Discovering that Pi was Christian, Muslim, and Hindu, each protested and demanded that Pi choose a single religion. Pi protested: "I just want to love God."

Science may be thought of as the search for God's eternal objects. Engineers apply that knowledge to the world; this can result in bombs (evil) or bridges (good).

Sacred texts also chronicle man's fallible search for God's eternal objects that have always existed, but continue to be revealed through God's possibilities in mankind's efforts in science, art, literature, music and religion. Like Pi, the religious must love God and, as with science, seek His truth.

Society

Society is supported and guided by religions.	
• Social Imaginary • Economics • Political • Socio-cultural • Humanism • Religious Suppliers • Societal Ethics	• How does social imaginary differ from theory? • How do primitive, traditional and capitalistic societies differ? • How do cultural, political & economic systems interact? • How did Western Societies evolve from Christianity? • What are some modern forms of western religions? • How has the church emerged over centuries? • How must religion/churches evolve to influence society?
Society is composed of communities that support the military, commerce, government and disciplines with socio-cultural norms for governance and citizen welfare. Societies evolve from their religious/cultural, economic and political systems. Whitehead asserts rational religion lies at the foundation of Western Society. Western society evolved from a theocracy to a secular democracy, which is being subjected to the detrimental systematic elimination of religion in the public sphere. Some would argue that proper behavior is innate in humans and that a society can provide a quality life for its citizens without God. This book agrees with the Roman Emperor Constantine and the USA's founding fathers: Christianity has and still provides a foundation for proper behavior and human values in Western society.	

Society should exist for the welfare of citizens. How will humans flourish? What is expected of citizens? How will citizens flourish? How will one govern? What socio-cultural norms are required for society to function?

Christianity provided a framework for Western societies that transformed their religious, cultural, economic and political systems to ones that are currently being undermined by secular humanism.

Since the beginning of recorded history, religion has been the primary force (other than military subjugation) for controlling tribes and societies by chiefs, medicine men, priests and kings. After many religious wars in Europe from 1524 CE to 1648, leaders and thinkers concluded the "ideal" government must separate church and state.

Religion and religious transformation provided the means by which theocracies evolved to secular democracies.

Social imaginary[15] is the way in which people imagine their social existence, how they fit together with others, how things go on between them and their fellows, the expectations, which are normally met, and the deeper notions and images, which underlie these expectations. Throughout history, religion has shaped social imaginary.

Social theory and imaginary differ in how:

1] ordinary people "imagine" and talk about society, not theoretically, but in images, stories, legends, etc.

2] a common understanding makes possible common practices, and a widely shared sense of legitimacy.

3] it is shared by large groups of people, if not all of society (social theory is often shared by a small minority).

In his book, *Legitimation Crisis*, Jürgen Habermas describes three types of societies:

- *Primitive* societies that are organized around kinship relations whose primary roles are determined by sex and age where there is no differentiation between social and system integration.
- *Traditional* societies that are organized by political classes, which are ruled by state power and socio-economic classes where their functional differentiation is between social and system integration.
- *Liberal capitalist* societies that are organized by unpolitical class rule, which is driven by labor and capital where an integrative economic system takes over socially integrative tasks.

---

[15] Social Imaginary and the history and evolution of Western Society is described in the book, *A Secular Age*, by Charles Taylor, Professor Emeritus of Philosophy at McGill University.

Habermas claims as they evolve, societies are always in danger of entering a state of crisis; crises arise when the structure of a social system cannot solve the problems required for its continued existence. These crises are not produced by external changes, but through internal incompatible system-imperatives that cannot be integrated. Crises disintegrate societies where their members feel their social identity threatened. He goes on to define:

- Social integration is related to the systems of institutions in which people are socially related;
- System integration is related to institutions with steering performances of self-regulated systems; and
- Social systems are life-worlds that are symbolically structured and maintain their boundaries and continued existence by mastering the complexity of a changing environment.

In modern capitalistic societies, the political system is subordinate to the socio-cultural and economic systems:

- Socio-cultural subsystems have standard structures that include a status system and sub-cultural forms of life with underlying categories that include distribution of privately available rewards and rights of disposition.
- Political subsystems have standard structures that include political institutions (state) with underlying categories that include distribution of legitimate power and available organizational rationality.

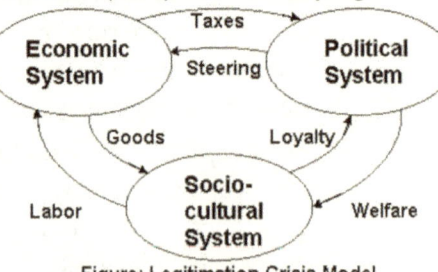

Figure: Legitimation Crisis Model

- Economic subsystems have standard structures that include economic institutions (relations of production) with underlying categories that include distribution of economic power and available forces of production.

Taylor's social imaginary alters Habermas' socio-cultural systems. Social imaginary extends beyond the immediate background of understanding, which makes sense of practices: one has a sense of how things usually are interwoven with how they are imagined and what missteps would invalidate the process. Consider the evolution in imaginary in the context of Habermas' societies:

1] In primitive societies, religious life was inseparably linked to social life. What stood out was the ubiquity of spirits, forces unlike the ordinary forces of animals; and how differently these higher forces related to society.

2] In traditional societies, higher religious forms appeared in 600 BCE lasting into the 16th century, marked by Confucius, Buddha, Socrates, and Hebrew prophets; they initiated a break in social order, cosmos, and human good. Humans were embedded in society, society in the cosmos and the cosmos in the divine.

3] In liberal capitalistic societies, modern humanism focused on human flourishing – prosperity, health, long life, fertility and the absence of disease, sterility, and death – involving no relation to anything higher. Secular society moved away from an unchallenged belief in God.

In 1500, Western Society made God's presence undeniable. In the old world:

- Cosmos was a hierarchy with higher and lower levels of being. Storms, droughts, floods, plagues, fertility and flourishing were seen as acts of God.
- Tension in earlier Christendom was between demands of self-transcendence and the requirements of ordinary human life. God was implicated in every aspect of society, which was interwoven with ritual and worship.
- Forces didn't just impinge on the individual, but on society in the enchanted world of spirits, demons, and moral forces; God's church defended the community.

- God's eternity according to Augustine made all time an instant of action; God is in the past, present and future. Meaning already existed outside of us (e.g. demons).

Today's exclusive humanism was a new sense of self, which was not open and porous and vulnerable to spirits and powers. In modern society:

- There is just linear time, a resource not to be wasted.
- Meaning comes into existence only as the world impinges on the mind/organism.
- The new cosmos is a universe of natural laws in linear time that are not evidently related to human meaning.
- Human minds composed of introspective self-awareness are the only minds in the cosmos.

Elites in history believed their way of life to be superior, but were content to exploit the lower orders until the 16th century. It became necessary to discipline the lower classes when rise in population and difficult economic times meant indigents increased, facilitating public disorder with increased crime and spread of disease.

Soon governments realized that as economic performance improved, tax revenues increased to support military power. As military technology advanced in the 17th century, and some states achieved greater advantage from production, governments became concerned with productivity. Many healthy and disciplined men were required to support production and the military.

Goals of civility and religious reform were frequently combined. Demands of civility for reordering society in turn gave a new social dimension to the pious, ordered life. While the Calvinist Reformation defined a path to Christian obedience, it also offered a solution to the social crisis of the age. During the 17th century, the state developed ordinances for the economic, educational, spiritual, and material well being on a wider stratum of their subjects. Effective government structures with the right spirit and discipline – supported by the Calvinist faith - were required to initiate social reforms. Modes of discipline, methods, and procedures proliferated; armies were primary, but some monies were applied to schools, hospitals and factories.

Locke's theory of natural law started with equal individuals in a society of mutual benefit. This turned into a society of the reconstruction. Rational, sociable beings, meant to live together with respect for each other's life and liberty, were also meant to preserve themselves by hard work. Western Christendom contributed both to the order implicit in intense piety, and the need to make God more fully present in every day life.

The newly remade society embodied the demands of the government in an uncompromising, stable, coherent, Modern Moral Order (MMO) with these crucial features:
1] It holds individuals together independent of hierarchy.
2] Benefits of and means to life relate to virtuous practice.
3] It secures freedom in terms of rights.
4] Rights, freedom and benefits are shared by all equally.

MMO gradually infiltrated and transformed the social imaginary. This involved people taking up, improvising or being inducted into new practices. It was eventually taken for granted.

Locke's society judged political performance by imagining social life independent of the political. Locke's moral order was crucial to represent a transformation of the social imaginary in: 1] practices and outlooks of democratic self-rule, 2] public sphere, 3] economy.

This expands upon Habermas' systems:

1] A political system of the "Sovereign People" was first legitimized in the U.S. Constitution. The American Revolution began with asserting "rights of Englishmen" against an arrogant, imperial government. Continental Congress reinterprets the existing Congregational church practice where the converted enjoyed full citizenship resulting in the U.S. constitution: "We, the people."

2] The public sphere, a component of the socio-cultural system, is new: a space grounded purely in its common actions where members of society meet - through print, media, face-to-face encounters (and now the Internet) – to discuss matters of interest to form a common mind. Freedom is essential to the rights that society exists to defend; government must win the consent of the governed to be legitimate. Before modern day, this was inconceivable except by something transcendent founded by God.

3] The economic system sanctifies ordinary life and promotes it as the highest form of Christian life.

Modern societies now look to exclusive humanism for their ethical foundations. However, for exclusive humanism to be more than a theory held by a tiny minority, but as a spiritual outlook, two conditions needed to be met: the enchanted world had to fade; and a viable conception of the highest spiritual and moral aspirations had to be possible without God.

Exclusive humanism seems unproblematic from within itself. Descartes, Hume, Locke claimed to be saying the obvious. Rather what happened is that experience was framed by a powerful theory, which posited primacy of individuals, the neutral, and the intra-mental as the locus of certainty. Values and virtues drove this theory: the independent, disengaged subject, reflexively controlling his own thought process. There is an ethic of independence, self-control, self-responsibility, and disengagement, which brings control; a stance which requires courage, the refusal of easy comforts of conformity to authority, and surrender to prompting of the senses. Emerging out of the careful, objective scrutiny, humanism is presented as being there from the beginning, driving the whole process of discovery, driven by its own "values."

Humanism's validity is bogus. What is passed off as simple discovery, is in fact a new construction; a change that involves a new sense of our identity and our place in the world, with its implicit values, rather than simply registering observable reality. It requires defense where there has been none.

Commerce and economies are also credited as a path to peace and orderly existence as opposed to a destructive aristocratic search for military glory. Adam Smith's 18th century theory conceived the economy as a system.

Today, the World Economic Forum (WEF) measures a nation's economic and political systems to determine its capability to create prosperity for its citizens. For over three decades, the WEF has studied and assessed many factors to provide insight on the best strategies to improve competitiveness. Competitiveness is a set of institutions, policies, and factors that determine the level of a nation's productivity. The level of productivity, in turn, sets a level of prosperity earned by an economy. The WEF 2016 report features 138 nations, and contains a profile for each economy and global rankings set by 114 indicators.

Since 2005, the WEF has based its analysis on the Global Competitiveness Index (GCI); its 12 'pillars' measure the microeconomic and macroeconomic foundations of national competitiveness. A more competitive economy is more likely to sustain growth. Consumption in today's global environment has made it important to account for social and environmental sustainability as well.

While all pillars matter, they affect economies in different ways. Because they're in different stages of development, improving Cambodia's competitiveness is not the same as improving France's competitiveness. The GCI assumes economies in the first stage are mainly factor-driven and compete based on their factor endowments—primarily low-skilled labor and natural resources. Companies compete on the basis of price and sell basic products or commodities, with their low productivity reflected in low wages. Maintaining competitiveness at this stage of development hinges primarily on:

Pillar 1: well-functioning public and private institutions,

Pillar 2: well-developed infrastructure,

Pillar 3: stable macroeconomic environment,

Pillar 4: healthy workforce with a basic education.

As countries become more competitive, they move into efficiency-driven development, where they must improve production processes and increase product quality because wages have risen and they cannot increase prices. At this point, competitiveness is increasingly driven by:

Pillar 5: higher education and training,

Pillar 6: efficient goods markets,

Pillar 7: well-functioning labor markets,

Pillar 8: developed financial markets,

Pillar 9: ability to benefit from existing technologies,

Pillar 10: a large domestic or foreign market.

Finally, as countries move into the innovation-driven stage, wages will have risen by so much that they are able to sustain them only if businesses are able to compete with new and/or unique products, services, models, and processes. At this stage, companies must compete by:

Pillar 11: sophisticated production processes and/or business models,

Pillar 12: producing new goods with new technologies.

Data on economic growth and employment show that, since World War II in Western economies, economic growth went hand in hand with improving living conditions, access to more and better goods and services for a growing portion of the population, and an overall enhancement of well-being. More recently the sharp rises in economic growth in developing and emerging markets have pulled hundreds of millions of people out of poverty, dramatically improving their living conditions.

However, the negative impacts of consumption associated with economic growth on the natural environment and on those unable to benefit from improving economic conditions raised concerns about sustainability in WEF's model. The model and rising population has brought about increasing pressure on water, energy, and mineral resources that are becoming scarcer as demand rises. The 2014 WEF report devoted 22 pages on how to "properly" understand and measure social and environmental sustainability to set policies. It proposed two more pillars (not listed in the 2016 report) for sustainable prosperity:

Pillar 13: Social sustainability: income inequality; youth unemployment; sanitation; water, healthcare and welfare.

Pillar 14: Environmental sustainability: ratify international environmental treaties to stringently enforce regulations for ecosystem protection, agricultural, water, fish stock and forest preservation, and CO_2 reduction.

Despite being ignored in the social sustainability pillar, what must religion's role be in WEF's economic world with a public sphere dominated by a cacophony of self-interest? Formulating and communicating ideals of universal peace, justice and well being for humanity and the planet and love of God would be a good place to start for an ecumenical theology and cooperating religions.

A universal ecumenical theology does not presume one ubiquitous religion. Today, there is a strong decline of religion in European societies, but not in the U.S.A. (see table) because:

1] American's social imaginary always saw itself as integrating differences, with "E pluribus Unum" as its motto. Unity was retained in a host of churches as a consensual "civil religion" of Protestant churches, but grew to include Catholics, Jews and Muslims.

Pew Forum on Religion & Public Life / U.S. Religious Landscape Survey 2006						
Americans are not dogmatic: % agreeing that...			Conception of God			
	% Many religions can lead to eternal life	% There is more than one way to interpret teachings of my religion	Net belief in God	Personal God	Impersonal God	Other/ Don't Know
	%	%	%	%	%	%
Total affiliated	68	68	92	60	25	7
Protestant	66	64	98	72	19	7
Evangelical churches	57	53	99	79	13	7
Mainline churches	83	82	97	62	26	8
Historically black churches	59	57	99	71	19	8
Catholic	79	77	97	60	29	8
Mormon	39	43	100	91	6	2
Jehovah's Witness	16	18	98	82	11	5
Orthodox	72	68	98	82	34	12
Jewish	82	89	83	25	50	8
Muslim	56	60*	92	41	42	10
Buddhist	86	90	75	20	45	10
Hindu	89	85	92	31	53	7
Unaffiliated			70	28	35	6
Atheist			21	6	12	3
Agnostic			55	14	36	5
Secular unaffiliated			66	20	40	7
Religious unaffiliated			94	49	35	9

*From "Muslim Americans: Middle Class and Mostly Mainstream," Pew Research Center, 2007.
Results based on those who are affiliated with a particular religion.

Throughout the report, figures may not add to 100 and nested figures may not add to the subtotal indicated due to rounding.

2] American and British intellectuals are both deeply invested in unbelief, but they are ignored in the U.S.A.

3] British religion followed ancient forms while the U.S.A. followed the protestant model from its inception.

4] Fusion of faith, family values and patriotism are still extremely important to American society; Right or Left.

5] For those seeking change, American pluralism affords them a model of numerous options of religious forms.

Rodney Stark defines a religious economy as all religious activity within a society: a market of organizations seeking to attract or maintain adherents to their religious cultures. His theory emphasizes the efforts of religious suppliers, while making assumptions about demand: individual capacities and preferences.

First, people are as rational (or irrational) in making religious choices as in making secular decisions. What people deem rewarding differs, being shaped by culture and socialization – laziness, impulsiveness and passion influence choices. Second, most behavior meets standards of rationality; studies reveal that religion can provide some people with good mental health – they are less prone to neuroses, anxiety, depression, and other psychosis. As for ignorance as a basis for faith, university faculty in the physical sciences is more religious than in other fields.

No supplier satisfies all niches in the religious market. In a religious pluralistic economy: 1] there are competitive efforts to appeal to each market niche; as a result, 2] public religious involvement is maximized; in addition, 3] effective organizations grow; ineffective ones decline.

Where religious monopolies prevail and are optional as in Europe, the overall level of public involvement is low. Unbelief and skepticism are not modern phenomenon. It requires vigorous effort to produce high levels of religious commitment on the part of the general public.

Taylor says spiritual quests today are a long-term trend, which has produced disciplined, committed individual believers, Lutherans, Calvinists, Methodists and today's pilgrim seekers, following their own path. He believes North Atlantic religion depends on the relations (hostile, indifferent, symbiotic), which develop between collective outcomes of a host of quests and centers of traditional religious authority between dwellers and seekers.

Within Habermas' political, socio-cultural and economic systems, Locke's moral order provides a foundation for understanding internal crisis in modern Western societies.

In 312 C.E., Constantine recognized the importance of belief in Christianity in governing the Roman Empire. For those who broke the law, eternal torment was a far more powerful deterrent than imprisonment. The Christian tenets for community, love of self, love of neighbor and love of God, also provided a guide for moral behavior and reduced the requirements for law enforcement.

The U.S.A.'s Founding Fathers recognized the same benefits from belief in Christianity as Constantine and their importance in creating a legitimate socio-cultural system. They overwhelmingly agreed that religion was crucial in sustaining a culture of responsibility needed to create a free society minimizing laws and taxes.[16]

The U.S.A.'s politicians and judges in the political system are passing new laws and reinterpreting old laws to reduce the impact of religion in society. As religious influence is reduced – by its failure to adapt to new revelation and because of the influence of secular humanism – more laws are being enacted to legislate proper/moral behavior.

This chapter emphasized the importance of authoritative religion in solving many problems in the socio-cultural system. Societal problems, in part, are not being solved due to religious rigidity and the influence of secular humanism on socio-cultural and political systems.

The previous chapter asserted that simple religious dogmas created for pragmatic reasons are dangerous guides. The spiritual quests of pilgrim seekers ought to be

---

[16] The first amendment states: "Congress shall make no law respecting the establishment of religion or prohibiting the free exercise thereof. " The Northwest Ordinance of 1787 emphasized the importance of religion in education: "Religion, morality and knowledge being necessary to the government of mankind, schools and means of education shall forever be encouraged."

The Declaration of Independence assumes God is sovereign in the affairs of society: "..all men are created equal, they are endowed by their Creator with certain inalienable rights...".

John Adams, not a devout Christian, stated: "We have no government armed with the power capable of contending with human passions unbridled by morality and religion."

validated by an evolving rational ecumenical theology supported by Whitehead's panentheism.[17]

Author's Reflections

Crimes committed in god's name pale in comparison to evil communist/atheist dictatorships. According to The Black Book of Communism, atheist leaders ordered over 130 million deaths. Not in war, but civilian slaughter; gulags and concentration camps; a bullet to the head; starvation - planned famines to punish dissidents (USSR 67 million, China 64 million).

The church defined Western society's 15th century value system. A great part of religion is played by imaginative presentation of spiritual truths. They are enshrined in modes of worship, religious literature, music and art. Religions cannot do without them, but if they dominate without criticism, they become idolatry.

Many ministers dwell on God's power and not truth of revelation. They see reason as contrary to belief in God and see their sacred texts as the inerrant word of God. Like Pi, most people just want to love God. Rather than loving God and seeking truth, they get trapped in religious traditions that dictate belief in their tenets and demands to support their traditions and leaders.

---

[17] Whitehead's Theory of Organism asserts that process alone is reality. Reality is an aggregate (nexus) of physical and mental events. God's primordial nature is the realm of eternal objects. Everything in the universe exhibits self-determination, not just humans. God's consequent nature influences free will, but not all occurrences are God's will (God is not evil). All experience (human, animal, atomic, and botanical) contributes to the ongoing process of reality. Using Whitehead's *panentheism*, Hartshorne's process theology explains human purpose and assumptions behind the sciences. Trusting in reason, process theology makes no appeal to revelation, faith, authority, mystic vision, paradox or other devices religious teachers use to exempt their ideas from criticism.

In the ancient world, religion dominated the socio-cultural system that influenced political and economic systems. Now socio-cultural systems in the public sphere dominate with media, universities, lobbyists, NGOs and religious communities. World Association of Non-Governmental Organization lists 22,875 NGOs (e.g., Green Peace, Save the Orcas, Save the Turtles, Salvation Army, etc.) claiming to be committed to universal peace, justice, and well being for all humanity. For 2016, opensecrets.org lists 11,106 lobbyists in Washington, D.C. who spent $3.15 billion attempting to influence US government legislation. Hartford Institute estimates 350,000 religious worship locations. There are over 4700 US degree-granting institutions and thousands of media outlets.

As a consequence, serious gaps caused by a cacophony of secular and religious beliefs include:

- Rituals: no longer reinforce societal order and law.
- Artifacts: are worshiped promoting idolatry rather than spirituality.
- Spirituality: relative morality predominates because of a belief in self or a weak belief in a god.
- Mysticism: unsubstantiated fantasy abounds; a rational view of God that embraces science is not ubiquitous.
- Myth: People believe science is truth rather than an approximation of the natural universe.
- Culture: celebrating diversity minimizes the positive impact of homogeneous mores requiring a proliferation of laws.
- History: Religious history and stories do not reflect the revelations of the last 500 years.
- Cosmology: Sacred texts contradict 21st century scientific approximations of the natural universe.
- Sacred texts: are worshiped rather than revised with new revelations to support science, society and law.

Although the world community will not accept ethical tenets from a major religion, it may accept a value system from a group representing all religions. An ecumenical theology must provide a framework that supports all major religions and the sciences. This perspective will also guide society's values and ethics and identify how each religion will nurture and interact in its society.

World religions evolved to contain their major elements: symbolic generalizations, beliefs, values, and examples. The ways that world religions indoctrinate the faithful in elements of the religion is a marvel. This includes stories, music, symbols, artifacts, education, creeds and prayers as well as spectacular shrines, temples, and churches. Places of worship were and are designed and built at great expense to create a feeling of God's presence.

Locke's 17th century economic system sanctified ordinary life promoting it as the highest form of Christian (Calvinist) life. Smith conceived the 18th century economy as a system of people acting in their individual economic self-interest. After decades of basing economic measures on these ideas, the World Economic Forum concluded it must (briefly) add social and environmental pillars.

An ecumenical theology must provide a framework that supports all major religions and the tenets of science. This perspective will also guide society's values and ethics and identify how each denomination will nurture and interact with the societies in which they are embedded.

The main tenets of Hindu saint Sri Ramakrishna's (1836:1886) doctrine of Ultimate Reality are:

- Ultimate Reality is the only one but has different names in different religions; it is personal as well as impersonal.
- Realization of the Ultimate Reality is the true goal and purpose of human life. It is the central purpose of all religions. Transcendent experience validates religions.

- There are several paths to the realization of Ultimate Reality. Each religion is such a path. As paths to the same ultimate goal, all world religions have validity.
- Each person should remain steadfast in his own path, without thinking that his path alone is true and perfect.
- One should respect founders of all religions as special manifestations of God and, knowing that God dwells in all people, one should serve all.

Swami Vivekananda (1863:1902) propounded this message at the Chicago Parliament of Religions. To the five principles of harmony, he added three corollaries:

- World Religions are complementary, not contradictory.
- There is no need to change one's religion for another.
- The ideal is to accept and assimilate the best elements of other religions while remaining steadfast in one's own.

Vedanta declares one Divine Truth but different illumined sages call the same Truth by different names that can be reached through different spiritual paths. When God's grace descends, each will understand his own mistake. Vedanta believes in inner spiritual transformation in individuals. It proposes to help Hindus become better Hindus, Buddhists to become better Buddhists and Christians to become better Christians. The seeker of Truth has a difficult task to perform, for he cannot affiliate himself to any particular branch of learning, but he cannot ignore the manifold character of knowledge.

The next chapter presents man's approximation to eternal objects in terms of the many disciplines that evolved over the last three centuries. Religion and its concept of God – aided by Whitehead – must include the revelations of these disciplines in its theology.

Knowledge

Knowledge is revealed faster than religions have adapted.	
• Disciplines	What disciplines innumerate mankind's knowledge?
• Science	What does one know from science?
• Evolution	Why is no god not provable?
• Atheism	What does one know from religion?
• Randomness	How must religion relate to all disciplines?
Communication technology, mathematics, engineering and the scientific method are translating theories into innovations at a spectacular rate. With the Internet and various other technological advances, a world community has been created with world-wide instantaneous communication. In the past, religion was integral with many bodies of knowledge such as: science, art, literature, philosophy, music, education, medicine, government, policy and law. Some would say that religion is out of touch with this modern reality. Since people's lives are value driven, religions must embrace a changing world with a universal concept of God and not simply counsel individuals to retreat into a 'personal' view of God.	

How did our intellectual disciplines evolve? Religion originally embraced the sciences. Flow of knowledge has always been part of human history, but the Greeks provided a foundation for western culture in 400 B.C.E. It then progressed slowly until 1500 C.E. After Descartes' philosophy, Newton's physics, Freud's psychology and Darwin's Theory of Evolution, and major advances in mathematics and engineering, knowledge expanded at a torrid pace. More was learned in the last century than in recorded history (useless narrative expands exponentially faster). Religion searched for truth in all things, but now derails the search for truth with belief in dogma.

Rodney Stark asserts real science arose only once: In Europe, and nowhere else. All societies had a highly developed alchemy and astrology, but only in Europe did these develop into chemistry and astronomy. Unlike most non-Christian religions, European theology assumed not only that the universe was created, but that its workings are logical and consistent, thereby being susceptible to reasoned inquiry. Consequently, science evolved only in Christian Europe primarily because only Europeans believed it could and should be done.

Whitehead grasped that Christian theology was essential for the rise of science, when he explained: "It must come from the medieval insistence on the rationality of God, conceived with the personal energy of Jehovah and with the rationality of a Greek philosopher. Every detail was supervised and ordered; the search into nature could only result in the vindication of the faith in rationality."

Major knowledge disciplines for the purpose of acquiring a graduate degree are listed in the table below:

Physical Sciences	Biological Sciences	Factual Social Sciences	Normative Social Sciences
• Astrophysics/Astronomy	• Ecology/Evolutionary Bio	• Anthropology	• Economics
• Chemistry	• Genetics & Genomics	• Communication	• Creative writing
• Computer Sciences	• Immun./Infectious Disease	• Geography	• Ethics
• Earth Sciences	• Kinesiology	• Linguistics	• Music composition
• Climate Sciences	• Microbiology	• Applied Psychology	• Political Science
• Physics	• Neuroscience/Neurobio	• Applied Sociology	• Philosophy
Engineering	• Physiology	**Arts and Humanities**	• Theology
• Aerospace Engineering	• Animal Sciences	• American Studies	**Religion**
• Biomedical Engineering	• Entomology	• Fine Arts	**Mathematics**
• Chemical Engineering	• Forestry/Forest Science	• Foreign Language	• Pure Mathematics
• Civil/Env'l Engineering	• Nutrition/ Food Science	• Literature	• Applied Mathematics
• Computer Engineering	• Plant Sciences	• History	• Statistics & Probability
• Electrical Engineering	• Agriculture	• Music	**Methodology**
• Engineering Science	**Medical Science**	• Theatre & Performance	• Scientific Method
• Materials Science	• Dentistry	• Film Studies	• Historical Method
• Mechanical Engineering	• Human Medicine/Nursing	**Government, Law, Policy**	• Theological Method
• Operations Research	• Pharmacy	**Business**	• Epistemology
• Architecture	• Public/Environ. Health	**Education**	• Theoretical Psychology
	• Veterinary Medicine	**Military**	• Theoretical Sociology

The reason science and mathematics are so important to the evolution of knowledge is the process by which their theories, facts, and knowledge are validated.

Testimonials in infomercials are either noise or information or lies but they are not knowledge.

A reporter must corroborate data by finding two independent sources to report her facts. If the reporter fancies herself a journalist, she theorizes to add her bias to the data without peer review by experts.

A lawyer through rhetoric and reference to precedent "proves" her point to a jury most of whom have no idea what correspondence, coherence or pragmatism mean in terms of validating her assertions.

Compare these to the scientific method. A scientist looks at disparate data or seemingly very different events and develops a theory that explains them. Her work is then submitted to the scientific community associated with that discipline with suggestions as to how to prove the theory WRONG. If, after the entire community cannot prove it wrong and provides additional proofs of its rightness, it is accepted and used by the community as *approximately* correct. This approach accelerated the progress of science because scientists trust their community's knowledge.[18]

By the end of the 19th century Charles Darwin's theory of evolution was established and by the end of the 20th was coupled with genetic theory. It is now used to explain the origin of all life and the universe. Evolution is based more or less on these five premises:

- The universe came into being out of nothingness approximately 14 billion years ago.
- Despite massive improbabilities, the properties of the universe appear to be precisely tuned to support life.
- While the mechanism of the origin of life on earth is unknown, once life arose, evolution and natural selection permitted development of biological diversity and complexity over very long time periods.
- Evolution required no special supernatural intervention since randomness is proposed as the agent of change.
- Humans are part of this process.

---

[18] Science is never absolutely right. For hundreds of years and from thousands of observations everyone knew that swans were white until one black swan was observed in Australia. A hundred successful experiments do not prove a theory right; one failed experiment proves a theory wrong.

Some biologists think this theory explains everything. Richard Dawkins, in his book *The God Delusion*, argues that the universe is explained by genetic theory and Darwin's theory of evolution and no God. Dawkins believes that the randomness at the genetic level is adequate to explain the entire formation of the Universe. Dawkins believes that billions of years of genetic replication, random mutation, and selection produced human beings with consciousness. Dawkins' argument is wrong. Randomness does not explain cause.

Assuming no randomness in the Universe implies that both the Theory of Evolution and Quantum Mechanics are inadequate. To be clear, there may be random events with no apparent prior cause; randomness successfully supports a theory of subatomic particle motion, but not the creation of the universe or evolution of life.

Niels Bohr, Albert Einstein and Richard Feynman, who all won a Nobel prize in Physics, wrote this about quantum theory: Bohr, "For those who are not shocked when they first come across quantum theory cannot possibly have understood it;" Einstein, "Quantum mechanics... says a lot, but does not really bring us any closer to the secret of the 'old one.' I, at any rate, am convinced that He does not throw dice." Feynman, "I think I can safely say that nobody understands quantum mechanics." Although physicists can explain subatomic phenomena, they don't know why randomness works!

Evolutionists believe natural selection can mimic the powers of design without itself being directed by any intelligence. Using genetic random mutation and natural selection in the last century has only managed to explain minor variations within a species. It has not established how one species can mutate into another or how simple organisms mutate into more complex ones. Unlike quantum physicists, evolutionists believe a theory that remains unproven after over 150 years.

Regrettably, evolutionists denounce anyone as IDiotic, ignorant, or a religious fanatic who suggests that random mutation and natural selection are inadequate to explain evolution. Darwinist professors deny tenure and seek to dismiss professors who propose Intelligent Design (ID). Also, challenging randomness in the universe poses a threat to atheists who use it to "prove" god is unnecessary.

In his book, The Edge of Evolution, Dr. Michael Behe observed that 10^{20} malaria cells were necessary to produce one mutation to combat curative effects of quinine (chloroquine). He concluded that if two such positive mutations were required to create a change in species, it would represent more animal species than have ever lived in Earth's history. Needless to say IDiots praised his brilliance while evolutionists proclaimed him delusional.

Some scientists like Behe seek evidence of intelligence in nature, showing that the complexity of human cells is not explained by random mutation and natural selection. In the last 25 years, scientists identified many examples of irreducible complexity (e.g., cellular flagellar motor, genes' machine code, physics' fine tuning, cosmology's goldilocks zone, fossils' Cambrian explosion).

Setting these arguments aside, consider that evolution is chaotic and not random. A random sequence of events is one in which anything that can ever happen, can happen next; like a coin flip; knowing the last toss is no help in predicting the next one. Conversely, chaos consists of events that aren't random, but seem to be: minor changes in initial conditions produce vastly different results. Chaos theory studies how unpredictably complex systems, once thought to be random, contain hidden ordered patterns.

In his book, A New Kind of Science, physicist Stephen Wolfram writes that his work on cellular automata showed that great complexity could be generated from simple programs. But, although he and others believed that his results were relevant to biological systems, the pervasive Darwinist belief is that level of complexity seen in biology must somehow only be associated with randomness and natural selection.

Wolfram does not challenge the idea of evolution as change over time, or common ancestry, but disputes Darwin's idea that the cause of all biological change is blind and undirected. Cellular automata or chaos theory may help establish algorithms (like RNA/DNA) to explain proliferation of species better than random mutation.

It is scientific to be skeptical about both entrenched Darwinism and Intelligent Design. Wolfram's cellular automata idea may better explain evolution, but if no one will admit that it is credible, Darwinism won't get "fixed." Any assertion must be established by scientific experiments that can be provably wrong, are subject to peer review and are critiqued in science journals.

This book adopts Whitehead's panentheism; it proposes that the universe is not random but chaotic and governed by laws of physics and biology and yet to be established algorithms.

No god is not provable, but many atheists have tried. The most coherent atheist publication is Physicist Victor J. Stenger's book, *God: The Failed Hypothesis*. He states that if one shows that particular attributes of God fail to agree with the data, then people would be irrational to use it as a guide for their religious activities. He defines God as a supreme being with the following attributes:

1] God is the creator and preserver of the universe.

2] God is the architect of the structure of the universe and the author of the laws of nature.

3] God is the source of morality and other human values such as freedom, justice and democracy.

4] God steps in whenever he wishes to change the course of events...

5] God creates and preserves life and humanity, where human beings are special in relation to other life forms.

6] God endowed humans with immaterial, eternal souls that... carry a person's character and selfhood.

7] God has revealed truths in scriptures and by communicating directly to selected individuals...

8] God does not deliberately hide from any human being who is open to finding evidence for his presence.

Stenger's scientific argument against the existence of God is a modified form of the lack-of-evidence argument:

- Hypothesize a God who plays an important role in the universe.
- Look for evidence of a God with specific attributes to provide objective evidence for his existence.
- If such evidence is found, conclude that God may exist; if it is not found conclude that a God with these properties does not exist.

Stenger summarizes each chapter in his conclusion, which is presented here with a hypothetical rebuttal. In accordance with Whitehead's view of God, the rebuttals presume God's consequent nature in influencing every event and His primordial nature embodied in eternal objects. This would remove from 1] 'creator', and change 4] to 'God participates in every event'.

The Illusion of Design: A God who is responsible for the complex structure of the world, especially living things, fails to agree with the empirical fact that this structure can be understood to arise from simple natural processes and shows none of the expected signs of design. Indeed, the universe looks as it should look in the absence of design.

Rebuttal: A God who is responsible for the complex structure of the world, especially living things, agrees with the empirical fact that very complex structures can be understood to arise from simple natural algorithms.

Indeed, the biologically complex structures that arose on earth from a vast empty universe affirm an elegant complex adaptive system.

Searching for a World beyond Matter: A God who has given humans immortal souls fails to agree with the empirical facts that human memories and personalities are determined by physical processes, that no nonphysical or extra-physical powers of the mind can be found, and that no evidence exists for an afterlife.

Rebuttal: A God, who gave humans the capacity for consciousness and self-awareness, did indeed give them souls. The ability to conceptualize and perceive oneself is the essence of soul whether in words (literature), motion (dance) or science (mathematics). The human soul rests in God's primordial nature (Whitehead): "my immediate occasion of experience, at the present moment, is only one among a stream of occasions which constitutes my soul."

Cosmic Evidence: A God whose interactions with humans, including miraculous interventions, reported in scriptures, is contradicted by the lack of independent evidence that these miraculous events took place and the fact that physical evidence so convincingly demonstrates that some of the most important biblical narratives, such as Exodus, never took place.

Rebuttal: A God who lures human action toward beauty and good is supported by the historical facts of sacred and secular texts. Some sacred writings are myths that convey revealing truths about culture; e.g. The Good Samaritan.

The Uncongenial Universe: A God who fine-tuned the laws and constants of physics for life, in particular human life, fails to agree with the fact that the universe is not congenial to human life, being tremendously wasteful of time, space, and matter from the human perspective. It also fails to agree with the fact that the universe is mostly composed of particles in random motion, with complex structures such as galaxies forming less than 4% of the mass and less than one particle out of billions.

Rebuttal: A God whose eternal objects are approximated by the laws of physics and genetics agrees with the fact that earth is congenial to human life, but not so congenial as to prevent evolution of culture and intellect and knowledge of God's intelligence in the universe as defined by mathematics of cosmology and the human genome. Dr. Francis Collins, head of the Human Genome Project, found "genomes to be elegant evidence of the relatedness of all living things and came to see this as the master plan of the same Almighty who caused the universe to come into being and set its physical parameters just precisely right to allow the creation of stars, planets, heavy elements and life itself."

The Failures of Revelation: A God who communicates directly with humans by means of revelation fails to agree with the fact that no claimed revelation has ever been confirmed empirically, while many have been falsified. No claimed revelation contains information that could not have been already in the person's head making the claim.

Rebuttal: A God who communicates with individuals by means of revelation agrees with the fact that science itself is represented by one amazing confirmed revelation after another; referring again to Whitehead: "(Science) must come from the medieval insistence on the rationality of God, conceived with the personal energy of Jehovah and with the rationality of a Greek philosopher."

Do Our Values Come from God?: A God who is the source of morality and human values does not exist since the evidence shows that humans define morals and values for themselves. This is not "relative morality." Believers and nonbelievers alike agree on a common set of morals and values. Even the most devout decide for themselves what is good and what is bad. Nonbelievers behave no less morally than believers.

Rebuttal: A God who is the source of morality and human values from the dawn of primitive man has always been imagined by man to exist and whose eternal objects and consequent nature have provided the foundation for all morality and human values. If nonbelievers are no less moral than believers, it shows that religious influence benefits all members of society.

The Argument from Evil: The existence of evil, in particular, gratuitous suffering, is logically inconsistent with an omniscient, omnibenevolent, omnipotent God (standard problem of evil).

Rebuttal: From the previous chapter, the universe is characterized by process and change carried out by the agents of free will. Self-determination is in everything, not just humans. God influences the exercise of free will by offering possibilities. God has a will in everything, but not everything that occurs is God's will (God is not evil).

These arguments for and against God only prove that a plausible narrative can be 'invented' for either viewpoint from the facts and are subject to narrative fallacy. Clearly, the scientific method is inadequate to prove or disprove the existence of God and that both Stenger's and the author's arguments are replete with fallacious narrative.

Physicist and Episcopal Minister John Polkinghorne, in his book *Faith of a Physicist*, points out that biologists like Dawkins suffer the same hubris as physicists in the 19th century with a theory of everything.

According to Vendanta, philosophy is illuminated by intuition and characterized by immediacy, universality, necessity and infallibility and perfect veracity. When it becomes impossible for reason to comprehend certain truths, it is not rational to reject super-rational experience as irrational. Concepts evolved from sense-experience are powerless in judging the nature of the ultimate Cause of all causes. No one can deny his own self or his being conscious of his self; nor can one deny that this consciousness is beyond the senses and reason. New knowledge may be had only in the intuition of Reality, when all knowledge is derived through the senses, understanding and reason falls short of it. No other method of approach to Truth than communion with the Absolute can give us reliable new knowledge.

At the beginning of the 20th century, Mathematician Bertrand Russell thought science had discovered all there was to discover: Mathematics, Newton's Laws of Physics, Einstein's Theory of Relativity, Quantum Mechanics, and The Big Bang framed our Universe comprehensively.

Consider, familiar things like suns, planets, gases and asteroids, etc., account for only 4% of the mass of the Universe.[19] Clueless cosmologists call the other 96% dark matter and dark energy. Their presence is determined by the rate of spin of spiral galaxies, like our Milky Way, which suggests it should be shedding stars (our Sun travels at 485,000 mph around our galaxy's core).

As the vastness of human ignorance unfolds, the God of the Gaps argument asserts: as more is learned, God's influence is reduced. Consider this analogy: Primitive man's points of knowledge on a straight line between zero and one might be ten points whereas our current knowledge is 10,000 points. Whether our points of knowledge are ten or 10,000, our points of ignorance are still infinite. God's primordial nature in his eternal objects is the infinite body of knowledge from which all disciplines are derived. Science in its ignorance postulates randomness.

---

[19] To validate this, George Smoot and John Mather at Berkeley designed the Cosmic Background Explorer (COBE) to look for extremely subtle differences in space that carry the imprint of the Universe when it was one second old. COBE found them: quantum fluctuations that 13.7 billion years later would coalesce into what is 22% dark matter, 74% dark energy and 4% of the known stuff which agrees with general relativity, but not quantum mechanics (ref: New York Times Magazine, 11Mar2007).

Noumenal Realm		
What we don't know we don't know >10^{10} more than we know		
?GOD?	**Phenomenal Realm**	? GOD?
	What we know we don't know >97%	
	What we know <3%	
What we know		**What we know we don't know**
• Tacit Knowledge		• Observed unexplained facts
• Artistic Knowledge		• Anomalies in theoretical knowledge
• Explicit Knowledge		• Perception of randomness
• Mathematics • Government • Agriculture		**What we don't know we don't know**
• Engineering • Science • Law		• Infinite amounts of data
• Humanities • Business • Military		• God's plan
Science explains < 0.1%		**Faith imagines the other > 99.9 %**

With all of our disciplines and knowledge, how much do we know that we don't know? The figure shows the relation of what we know and don't know. It is kinder than the previous analogy and suggests that there is only 10^{10} times more knowledge to acquire. Science progresses by imagining previously unimaginable realities. Great science does more than validate ideas; it predicts something that has not been observed. [20]

In 1931 Kurt Gödel proved that all mathematical areas contain true statements, which are NOT provable. Mathematics, often referred to as the Queen of Sciences, will always be fuzzy around the edges; demonstrating the fundamental nature of scientific limitations.

Kant distinguished between two realms of reality: phenomenal and noumenal. Phenomenal pertains to the world of appearances, whereas noumenal is simply the realm of the thing-in-itself. Man knows the phenomenal realm, while the noumenal realm remains hidden and beyond man's comprehension. However, transcendent reason takes man

---

[20] E.g., Einstein predicted light would be bent by a large mass and space-time would be curved. He had not observed this. His famous $E = mc^2$ was just taking $E=mv^2$ and pushing the velocity to the speed of light thereby showing that Energy is nothing other than mass times a conversion factor. In the case of the atom, it happens to release fantastic energy. Without observing this, he predicted it.

beyond observation to speculate about the noumenal realm and can expand his phenomenal realm of reality.

Mans' revelations continue to expand science into the noumenal realm of reality. While proving God's existence may be beyond the proper realm of phenomenal reality, science looks for and finds order in our universe. While not proving His existence, science assumes God's existence as it continues to look for more rational order.

God is not definable, but a concept of God is foundational in developing a framework to harmonize the physical and social sciences and to derive values for communities, individuals and society. If we are indeed created in God's image neuro-physiologically perhaps we can lend some detail to a panentheistic view of the mind of God and His relationship to our universe and us (illustrated later).

Author's Imagining: A Social Systems Model

Without theory, experience has no meaning;
Without theory, one has no questions to ask;
Without theory, there is no learning.
- Dr. W. Edwards Deming's Theory of Knowledge

Atheists argue that our thinking is now so knowledgeable and erudite that god is no longer necessary to support an explanation of our relationships, nature, the planet, the universe and ourselves. One must adamantly ignore things that we know that we don't know to be this certain about no god. It is a sad commentary on intellectuals that conclude: "there is no god."

To refute this narrative, we will take a "systems" view of God. As a Systems Engineer by profession, philosophy and apologetics, and social sciences seem unsatisfactory since their primary objects are words with weak structural models. Therefore, a social systems model is presented here to provide a framework to illustrate relationships in economic, political, and socio-cultural systems.

Encyclopedia of the Social Sciences defines: general systems theory as a program (not a scientific theory) in the contemporary philosophy of science. All variants have a common aim: the integration of diverse content areas by means of a unified conceptual methodology.

System refers to widely separated concepts. Engineers see it as aggregates of technological devices. Physiologists define functionally related portions of living organisms (circulatory, digestive, nervous systems). Social scientists speak of economic and political systems and philosophers about systems of thought. A system can be defined as:
1] something consisting of a set of entities
2] among which a set of relations is specified, so that
3] deductions are possible between relations or from relations among entities to system behavior or history.

In cosmology, a solar system contains a sun and planets; relations among them specify position and velocity vectors and forces of gravitational attraction. In language, phonemes, morphemes, sentences, etc. and relations between them are syntactic and semantic rules.

In systems engineering over the past forty years using a cybernetic multi-level isomorphism model,[21] Stafford Beer developed a "social systems" view of individual, community, organization and society in a business environment (*ref: Brain of the Firm*), shown in the figure.

- System 5 = Cortex: Higher Brain functions
  = Organizational identity; ultimate control

---

[21] Cybernetics (Merriam-Webster dictionary) science is concerned with the comparative study of control systems (e,g,, the nervous system and brain and electrical communication systems). Two objects are isomorphic if there is a one-to-one correspondence between the elements of one and those of the other and if relations among the elements are preserved by the same correspondence; e.g., a mechanical harmonic oscillator and an electrical circuit. Multilevel implies levels of the organizational hierarchy.

- System 4 = Diencephalon: Input from senses = Environmental scanning; forward planning; adaptation
- System 3 = Base Brain: Pons and Medulla = Internal regulation; optimization; synergy
- System 2 = Sympathetic = Nervous System: Stabilize muscles & organs = Coordination; conflict resolution
- System 1 = Muscles and organs: Maintain human body = Operations; primary activities

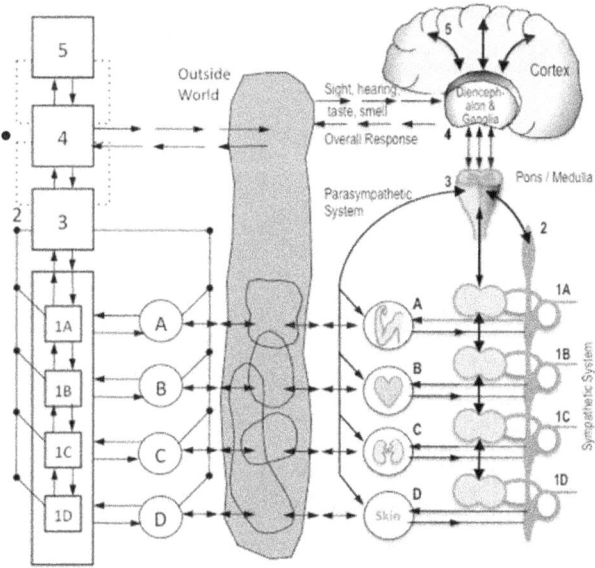

The systems of a firm having subsidiaries A, B, C, & D

Two dimensions of neurophysiological control: the main vertical command system (somatic) and the sympathetic and parasympathetic system (autonomic)

He proposed that for an organization to be viable, each level of its management hierarchy must be structured in such a way that it mimics the human nervous system.

Now consider Beer's Viable Systems Model by reworking the organization part of the figure where:

- Viability Systems' key proposition is that policy, intelligence, cohesion, coordination and implementation are distributed at all levels of the organization.
- System 1: Implementations produce an organization's purposes.

- System 2: Coordination provides capabilities that support System 1 activities.
- System 3: Cohesion steers the implementation functions.
- System 4: Intelligence proposes how to adapt to the external environment.
- System 5: Policy designs debates with balanced contributions of the cohesion and intelligence functions, which offer alternative perspectives on the shared adaptation to problems.

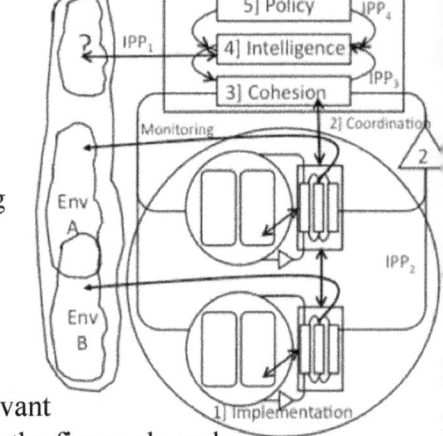

Policy making requires the orchestration and monitoring of organizational debates to enable people to argue for organizational adaptation and survival.

Acquiring and reporting relevant information is represented in the figure above by

- **IPP$_1$** – identify problems in the external environment requiring changes to internal activities to produce new products, services or behaviors.
- **IPP$_2$** – coordinate and enable production (internal) activities cohesively.
- **IPP$_3$** - support adaptation by communicating between the Cohesion and Intelligence functions.
- **IPP$_4$** - monitor and summarize IPP$_3$ for effective policy.

Notice that system '1s' have the same management structure as the next level of the organizational hierarchy. This recursion can be repeated for subsidiary, company, market and government; boxes are shorthand for each level represented in the next figure. Consider what information must flow by expanding the WEF's economic model to include the political and socio-cultural systems:

- **IPP$_1$** – identify novel ideas and solutions for a problematic economic and ecological environment.
- **IPP$_2$** – enable cohesive inter-societal activities.
- **IPP$_3$** - support communication between the Cohesion (3) and Intelligence functions (4) using the UN, WEF, Pontifical Academy of Sciences and an ecumenical theology to support cooperation and discourage conflict.
- **IPP$_4$** - monitor and summarize IPP$_3$ for effective policy making between nations and visioning to harmonize planetary ecological and economic activities (e.g. WEF's Global Competitive Index).

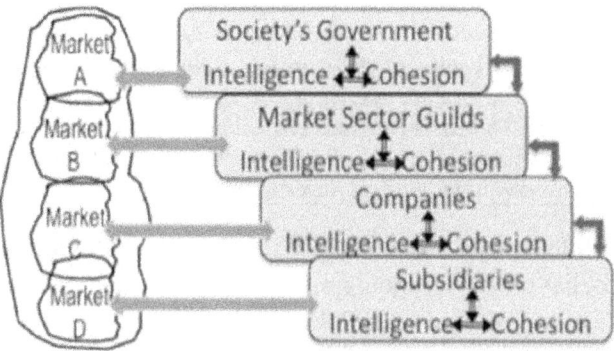

4 Level Hierarchy of a Society's Economy

The genius of Beer's view is that it provides a general and uniform way of looking at all levels of a "conscious system." A similar social systems model will be developed in the next chapter for Individual > Community > Society > Planet > Universe > God and will speculate on what flows on the associated IPP$_1$ through IPP$_4$.

Theology

Theology must guide religions to a universal value system.	
• Philosophy • Quality / ideas • Quantum Theory • Berkeley, Pirsig, James, Northrop, Whitehead	• *What is reality?* • *What is static & dynamic quality/value/knowledge?* • *What are levels of consciousness?* • *What are the stages of science?* • *What are the elements of value?* • *What might the mind of God be like?*
Human life, institutions and all disciplines are driven by value, which science excludes. Whitehead's Theory of Organism provides a foundation for bringing philosophy, science, mathematics, art, literature and religion under one framework. God's primordial and consequent nature elegantly explains reality and has been adopted by theologians of all world religions. **Two thousand years ago theology was a continuously evolving search for God's truth, to explain everything. Religions must return to that tradition** *within a collaborative religious environment to: improve human welfare, encourage harmonious beliefs of disparate religious views and provide collective and individual guidance to people.*	

Philosophy rationalizes what is known with what is believed. What should a modern theology look like? How would it define value? How can this theology be used to define the elements of value?

The previous chapter on knowledge argued that the universe is elegantly ordered, and there is no supernatural and no randomness only limited perception of natural law.

A theological framework must be developed for the clergy of all religions to discuss the critical values and ethics in societies with diverse religious beliefs and socio-cultural, economic and political systems. It must embrace the disciplines in modern academic thought. For those who object to the word religion, consider Albert Einstein: "I have not found a better expression than 'religious' for the trust in the rational nature of reality that is, at least to a certain extent, accessible to human reason."

Most religions acknowledge that God is not imaginable; nevertheless, they persist in imagining God. Whitehead's primordial and consequent nature of God represents a subset of the characteristics assigned to God in most religions. Representing 80% of the world's population, Christianity's Trinity, Hinduism's Trinity, Buddhism's indescribable God, and Islam's unknowable God are man's flawed attempts to imagine God.

Berkeley's ideas, Pirsig's values, and Whitehead's process as reality along with William James' and Northrop's philosophies provide background to frame a theology and define the elements of value in the context of God's continuous revelation.

Bishop George Berkeley (1685-1753) avoided the scientific subject/object metaphysics of Descartes and Newton. His idealism is simply stated in Latin as "esse est percipi" (to exist is to be perceived).[22]

Berkeley attacks the scientific assumption that material things are mind-independent things or substances that would exist whether or not any thinking things (minds) existed. Berkeley's argument is simplified as follows:
1] One perceives ordinary objects (rocks, birds, etc.).
2] One perceives only ideas. Therefore,
3] Ordinary objects are ideas.

In addition to perceived things (ideas), he imagines perceivers, i.e., minds or spirits, as he often terms them. Spirits, he emphasizes, are totally different in kind from ideas, for they are active where ideas are passive. God is an infinite spirit. Berkeley argues for God's existence as the cause of our sensory ideas by elimination: What could

---

[22] From his lecture at the City College of New York, Darren Stalaff outlines Berkeley's theory and book - The Principles of Human Knowledge: 1] A consistent empiricism concludes that the only known phenomena are ideas. Materialism and matter are dogmatic superstitions. 2] Newtonian physics shows us the law-like regularity in the sense impressions received from God. 3] Belief in matter is the cause of skepticism because one is separated from the "material" world by a veil of ideas that can never be rendered.

cause my sensory ideas, supposing that matter doesn't exist?: 1] other ideas, 2] myself, 3] some other spirit.

According to Krishnananda, objects do not arise from an imaginary abstraction. To the Vedanta, nothing appears without a substratum of reality for it is a necessary implication of our existence. The Vedanta, however, would accept Berkeley's position that other minds perceiving the same objects proves not the independent existence of objects, but that all minds are limited to a similar perception. In the Vedanta, however, matter is outside any particular individual mind, though it loses its materiality when it becomes the content of minds. Both agree that the universe exists in the mind of God, but Berkeley's 'imaginary;' is Vedanta's 'false perception'.

Newtonian physics presumed to discredit Berkeley's subjective idealism until 20th century Quantum Theory. The interconnectedness of everything at the quantum level, leads to the view that God, the universe, and people are all connected. The particles that make up people are connected to everywhere in the universe; there is no quantum reality independent of observation: "esse est percipi."

Whitehead partially based his theory of organism on Berkeley's philosophy that a "veil of ideas" separate us from experiencing reality. Whitehead resolves the issues and quiets Newtonian criticism with his theory of perception presented earlier. Whitehead's novel contribution to philosophy is that ALL reality is based on process and not on substance (objects).

Krishnananda explains that Whitehead's God may be regarded as an expression of the impulse to advance in consciousness and evolution. The Vedanta identifies his subjective aim with the aspiration of the universe to realize its perfection in the Absolute, which is immanent in the actual occasions and the eternal objects. Whitehead holds that God and the universe are mutually immanent and interpenetrative. Whitehead's God is analogous to

Vedantins' Isvara, who is accidental in the sense that He relates to the particular universe of which he is Lord; God then becomes 'consequent' related to the evolving process, and so is a manifestation of the Absolute.

Krishnananda surmises the problem of causation is not solved by process. Process is the relative appearance of a more fundamental reality. The Vedanta identifies this reality with the immutable consciousness immanent in all processes and yet transcending them. Whitehead's process may be real to us, but is not real in itself. Matter, life and mind step beyond their distinctness of structure and realize themselves in the truth of the Absolute.

When he argues that the universe as process is false, Krishnananda misinterprets Whitehead's theory; universe as process does a brilliant job of explaining the waking (Stage 1) and dream state (Stage 2). Eternal objects explain Vedanta's Reality in Isvara (Stage 3). Brahman (Stage 4) is beyond comprehension. Eastern philosophy has found in Whitehead's universe as process and eternal objects, a philosophy that explains the three stages of the Atman of Indian philosophy eloquently, elegantly, usefully, scientifically, and specifically.

In his book, *Lila, An Inquiry into Morals*, Robert Pirsig also relates his Metaphysics of Quality (MOQ) to a probabilistic view of Quantum Theory. Since quantum subatomic particles are "probabilistic" waves or particles that compose everything there is, like Whitehead, he concludes, that there is no substance in the world. Quality is the world's primary empirical reality. Quality is morality (and value). The idea of a rock is really quantum particles in constant motion, not unchanging substance.

Empiricism claims that legitimate human knowledge arises from the senses and denies the validity of knowledge gained through imagination, authority, tradition, or theory. MOQ says these are valid, and have been excluded by a metaphysical assumption that anything that can't be classified as a subject or object isn't real.

In MOQ, "causation" can be replaced by "value." To say that "A causes B" or to say that "B values precondition A" is to say the same thing. 'Substance' is replaced by 'stable pattern of value' or Berkeley's "law-like regularity in sense impressions received from God."

According to Pirsig, each culture has its own pattern of static quality derived from its fixed laws, traditions and values. But, in addition, there is a dynamic quality (revelation)[23] that is outside of the culture's existing system of precepts, but has to be continually discovered as culture evolves. Using Kuhn's Scientific Community or trial and error provides the necessary stabilizing force to protect delusional babbling from degenerating into cultural chaos.

Pirsig divides static patterns of value into four discrete systems: inorganic, biological, social, and intellectual. Although each higher level is built on a lower one, it is not an extension of that lower level. The higher level is seen as controlling the lower level where possible for its own purposes (e.g., consider writing this book on a computer where there is a hardware level, system level and word processing software). This is impossible in a Subject-Object Metaphysics (SOM) where everything has to be an extension of matter. But now atoms and molecules are just one of four levels of static patterns of quality and there is no intellectual requirement that any

---

[23] Dynamic quality emerges from intuition and immediately apprehended fact. Beneditto Croce in Aesthetics writes: impressionistic art "is intuition, in so far as it is a mode of knowledge, not abstract, but concrete, and in so far as it uses the real, without changing or falsifying it." Conveying immediately apprehended facts restricts one to impressionist art with nothing but its sense impressions. This chapter references essays on art, literature and science. Science cannot be used alone to define and criticize human activity and God, because science doesn't speak to the issue of value, which is the primary driving force for humanity.

level dominate the other three. Value holding a glass of water together is an inorganic pattern of value. Value holding a nation together is a social pattern of value.

Pirsig asserts that what is at a higher level of evolution is more moral. Some might disagree since putting socio-economic systems above eco-systems is making vast areas of the planet uninhabitable.

Static quality is his elementary unit of ethics upon which all right and wrong is based. MOQ says there is a morality called "laws of nature" by which inorganic patterns triumph over chaos; there is a morality called "law of the jungle" where biology triumphs over inorganic forces of starvation and death; the law is a morality where social patterns triumph over biology; there is an intellectual morality that attempts to understand and control society. Dynamic quality in terms of religion is God's continuing revelation and static quality is man's interpretation of it into religious and scientific laws. To Pirsig everything is an ethical process.

Whitehead's process defines value as subjective aim: the ideal of what a subject could become shapes the very nature of the process result. God through eternal objects offers for each actual occasion of process, a vision of what it might become.

In Whitehead's view, the higher the level of consciousness of the prehending subject, the more influence it will have on the outcome. Whitehead's theory of organism states that everything changes. Rocks maintain their rockness because their low level of consciousness assures that they follow physical laws of nature with little creativity despite frantic activity at the quantum level.

Subjects with higher levels of consciousness cause much higher variation in the processes in which they participate. Animals and plants exhibit consciousness in Whitehead's theory, but exert less variation than human consciousness.

Evil is not introduced into the universe by God, but by other forms of consciousness (e.g. felons, leaders, bullies).

Swami Sivananda's system of philosophy perceives that the basis of all knowledge and other ways of knowing are meaningful by illumination of the light from the Absolute.[24] The philosophy of the Absolute rises above particulars to greater universals, basing itself on facts of observation and experience by the method of induction and gradual generalization of truths, without missing a link in the chain of logic and argumentation, reflection and contemplation, until it reaches the highest generalization of the Absolute Truth; and then by the deductive method comes down to interpret and explain the facts of experience in the nature of this Truth.

To understand how science absorbs dynamic quality into static quality, consider Yale Professor F. S. C. Northrop's ideas from: *The Logic of the Sciences and Humanities.*

The first stage of scientific inquiry is what one directly perceives: relevant facts from a stream of consciousness when one is awake. Pure fact (dynamic quality) is a continuum of aesthetic qualities, not an external material object. Scientific objects are theoretical objects; not immediately apprehended facts. Scientific knowledge that requires an objective public world with scientific objects in it to be the same for all observers, is theoretically inferred, not empirically given knowledge. All one knows as pure fact (as with Berkeley) is what one's senses convey that is neither material common-sense nor scientific objects, but intermittent aesthetic qualities for

---

[24] Beauty is the vision of the Absolute through the senses. The beauty of symmetry, rhythm, harmony, equilibrium, and unity is manifested in consciousness. Aesthetic consciousness is thus the result of a partial expression of universal in conscious experience. All intrinsic or extrinsic values are rooted in the judgment of the supreme value of realizing the Absolute, where values are fulfilled.

each person; they convey neither substance nor causality in the sense of a public world.

The second stage of scientific inquiry are concepts of intuition whose complete meaning must be found in factors, which can be immediately apprehended in the modes of causal efficacy and presentational immediacy. The second stage of the scientific method inspects the relevant facts designated by the first stage of inquiry. The second stage begins with immediately apprehended fact and ends with described fact in the mode of symbolic reference. Since methods of the second stage of inquiry[25] are inductive involving 1] observation, 2] description and 3] classification, concepts are descriptive and qualitative in character. They are concepts that are immediately apprehended. Biology with its classification of genera and species constructed in terms of directly observable characteristics is a science in the second stage.

Unobservable objects of the third stage are concepts by postulation or intuition. Their existence is constructed in a deductively formulated system that designates explicitly what is proposed to exist. To this hypothesis, formal logic deduces theorems that define experiments to be performed. If all experiments give the result called for by the theorems, the hypothesis is confirmed and its entities and relations are said to exist (E.g., Physics, Engineering).

MOQ's static quality, as represented by knowledge in graduate programs, and Northrop's method of scientific inquiry are illustrated on the table on the next page. Since an individual must study for more than seven years beyond high school to attain a PhD in these academic topics, it could be argued that they represent a partial list of societal values.

---

[25] As one learns, one retains in memory more objects for symbolic reference (classify) that influence the field of consciousness in the mode of causal efficacy (describe), which in turn expands prehensions (observe) in the mode of presentational immediacy.

Physical	Biological	Factual Social	Normative Social
• Astrophysics/Astronomy	• Ecology/Evolutionary Bio	• Anthropology	• Economics
• Chemistry	• Genetics & Genomics	• Communication	• Creative writing
• Computer Sciences	• Immun./ Disease	• Geography	• Ethics
• Earth Sciences	• Kinesiology	• Linguistics	• Music composition
• Climate Sciences	• Microbiology	• Applied Psychology	• Political Science
• Physics	• Neuroscience/Neurobio	• Applied Sociology	• Philosophy
Engineering	• Physiology	**Arts and Humanities**	• Theology
• Aerospace Engineering	• Animal Sciences	• American Studies	**Religion**
• Biomedical Engineering	• Entomology	• Fine Arts	**Mathematics**
• Chemical Engineering	• Forestry/Forest Science	• Foreign Language	• Pure Mathematics
• Civil/Env'l Engineering	• Nutrition/ Food Science	• Literature	• Applied Mathematics
• Computer Engineering	• Plant Sciences	• History	• Statistics & Probability
• Electrical Engineering	• Agriculture	• Music	**Methodology**
• Engineering Science	**Medical Science**	• Theatre & Performance	• Scientific Method
• Materials Science	• Dentistry	• Film Studies	• Historical Method
• Mechanical Engineering	• Human Medicine/Nursing	**Government, Law, Policy**	• Theological Method
• Operations Research	• Pharmacy	**Business**	• Epistemology
• Architecture	• Public/Environ. Health	**Education**	• Theoretical Psychology
	• Veterinary Medicine	**Military**	• Theoretical Sociology

		Intellectual Static Quality: What one perceives to have been.
Third Stage of Scientific Inquiry	Concepts by Postulation	Unobservable objects existence is posited by a deductively formulated system whose basic assumptions or postulates designate unambiguously what is proposed to exist.
Second Stage of Scientific Inquiry	Concepts of Intuition	Ideas designate the inner objects of contemplation, whether these be individual things, like 'the sun' or classes of things like 'animal kingdom' or 'rationality' (from James). Synthesis of relevant facts.
First Stage of Scientific Inquiry	What one directly perceives	Relevant facts from a Stream of consciousness that is always going on when one is awake (James).
		Dynamic Quality: What is (Reality) in the moment.

An explicitly defined view of static quality, derived from Pirsig's "social" is divided into factual and normative social sciences. Intellectual static quality is what one perceives to have been (waking and dream states) where intellectual dynamic quality (Reality) is in the moment (deep sleep and transcendent states).

Earlier, social imaginary was defined as the way in which people imagine their social existence and that a chaotic value system was the result of over 130,000 groups influencing the USA's public sphere. Clearly, an authoritative value system needs to be defined to quiet the chaos and positively influence social imaginary.

Religious misuse of authority is well documented, but consider the misuse of ecological science in the first stage of scientific inquiry; the results of this research is treated by some as if it was in its third stage of inquiry to influence social imaginary and environmental regulation whereas, in reality, experts are guessing.

Protecting ecological resources must consider WEF's economic factors to improve human welfare. Environmental regulation based on educated guesses from ecological science and overestimating the ability of new technologies to produce adequate energy can have a significant negative impact on nations' energy production, which negatively impacts economic factors and citizens' welfare.

Religion must support the synthesis of scientific, ecological, and economic factors to authoritatively determine societal value that guides public policy, economic welfare, and cultural harmony.

Civilization may depend on the faith of men in beliefs, which can be justified and preserved only by methods of the third stage of scientific inquiry. They require one to think deductively, and have a greater relevance for the humanities, religion and problems of value than does the second stage of inductive scientific inquiry.

What Pirsig calls "static intellectual" applies to the four static categories – inorganic, biological, social, and intellectual - and represents our current limited knowledge of universal laws. As more of the four systems "laws" are revealed to us by divine or human revelation, community (Kuhn) must decide, within the four systems, if they are indeed a "better" view or nonsense.

Morality defined from a theology, which would guide science must connect itself to contemporary science. Catholicism through St. Thomas Aquinas is connected intimately with the science of Aristotle, but not that of Galileo, Newton, Darwin and Einstein. Protestant doctrine treats religion as independent of natural science. It provides a subjective personal ethic, but is inadequate to generate an ethical societal framework to control stem cell research, nuclear warfare, and cloning. Before one thing can hope to influence another it must connect to it.

With Aquinas, religion sought "God's Laws" for Pirsig's four major static categories. Physical and biological sciences have evolved rapidly with mathematics, physics and evolution. Religion, sociology, theology, psychology and philosophy have failed to create universal societal values. Once one "imagines" that immutable "laws" govern society, they will be found.

A theology for a religion to develop values is not just to guide the faithful, but to provide a universal interpretation of "right action" for the socio-cultural and the political systems that provide the greatest good for the greatest number of people. Engineering justice from values into law must be subject to peer review.

Ottowa University's Associate Professor of Philosophy Andrew Sneddon synthesizes Whitehead's and Pirsig's philosophies of value into components;

A] Repetition: Reiteration of a pattern in the actual world is endurance of form. Whitehead sees this as derived from eternal objects in God's primordial nature. Pirsig's dynamic patterns are preserved if they are felt to be successful.

B] Novelty: Mere repetition is not an ideal value state. Whitehead and Pirsig see 'life' as the embodied impulse towards novelty away from stale patterns of existence.

C] Definition: Value is the outcome of limitation. Both Pirsig and Whitehead recognize that 'decisions' are rated by: internal 'depth' of the satisfaction and social utility.

D] Contrast: Pirsig's dynamic quality is shaped into static patterns; higher patterns are different from lower patterns, but require them in order to come into existence. Whitehead's higher contrasts require lesser ones. Process-reality evolves--occasions depend upon past forms of experience to position them for new, higher levels.

E] Limitation: Value experience necessarily involves limitation--differentiating one thing from another, or favoring one and rejecting the second.

F] Final Causation: Value is not vague and should be used for primary classification; it involves proposing an end to achieve. Nothing is vague about a citizen voting.

G] World-Oriented: The more one positively charges this world with value, the better the world, and the more developed one's character. Value-reality exhorts each individual to conduct himself to achieve the greater good.

The objective of a theological community should be to restore the authority of religion in the society and charge it with value that limits and harmonizes human conduct.

Social scientists treat religion as a universal, but they adopt a low concept of it, based on primitive rituals, mythology, institutions, etc. Denying God's primordial and consequent nature will assure social science of continued failure. That humanity and not God are imminent in the Universe is hubris. However, skepticism about God's nature and participation is necessary for continued revelation and understanding of self, society, disciplines, and God.

Hindu Vivekananda's high concept of religion identifies it with man's struggle to attain transcendental spiritual consciousness and experience. Not part of any religion, Universal Religion is humanity's universal spiritual consciousness and heritage.

Constantine and the USA's founding fathers understood that religion positively charged society with value. It permitted a free society because it reduced requirements and expenses for law enforcement and imprisonment.

Each person is responsible for their individual conduct recognizing she/he is a moment of value activity that composes the universe as described by Whitehead's universal relatedness. To limit a 'self-interest' theory to a narrow, 'survival of the fittest' ignores obvious evidence. Individuals' interests are related to their value of self and how they care for others and the planet.

Sacred texts represent Pirsig's static quality for the socio-cultural system and normative social science. But when their adherents think they know the truth and stop seeking it, in Karen Armstrong's words, "they become a toxic arsenal that fuels hatred and sterile polemic."

Faith, transcendence, perception and imagination are part of the human consciousness, which is the path to ALL knowledge (and God's eternal objects): Scientific inquiry in its first, second and third stages establishes whether new information is valid, invalid or indeterminate, which may require an imaginative application of a new social systems theory. Questions about ethics and value and dynamic quality may be illuminated with imaginative application of cybernetic models.

Author's Imagining: Domains of Consciousness

Some intellectuals (?) support attempts to prove no god with the fallacious narrative that value can be established "scientifically." Because God (as does infinity) and value defy a logical explanation, does not mean that the concept of God is irrational.

Determining value is not the purpose of science, but its method can be used by a theology to partially validate a rational religion. God is fundamental and necessary for a theology to understand the value and purpose of the individual, community, society, evolution of knowledge and science. This theology will provide a foundation for psychology, sociology and societal ethics. To accomplish this, however, it must consider our level of connectedness throughout the planet.

Some claim that as the Internet makes information more available, an illusion of knowledge is fostered, resulting in individuals becoming dumber; a Pew survey concludes:

- Google will not make us stupid.
- By 2020, Internet use will boost human intelligence as people become smarter and make better choices as a result of unprecedented access to information.
- The Internet will help to improve reading, writing, and the rendering of knowledge.

Princeton's Global Consciousness Project (GCP) created a technical structure that "may have the capacity to register evidence of interactive connections of minds on a world-spanning scale." Testing over ten years found: "anomalous trends in the data is evidence that a global consciousness exists in a faint but detectable form."

In an interview Christof Koch, Professor of Cognitive and Behavioral Biology at Cal Tech, said: "...the Internet now already has a couple of billion nodes. Each node is a computer. Each one of these computers contains a couple of billion transistors, so it is in principle possible that the complexity of the Internet is... conscious..."

When asked why his book quoted Jesuit Priest Pierre Teilhard de Chardin (mentioned earlier), Koch replied: "...he argued that from very simple micro molecules to single cell organisms to multi-cell organisms to simple animals to complex animals to us is the emergence of complexity. He observed that the universe was getting more and more complex... Essentially, he postulated something like the Internet. He called it the "noosphere" -- the sphere of knowledge that covers the entire planet and is heavily interconnected. He died in 1955, long before any of this emerged, and he postulated that human society would evolve... (to be) self-conscious..."

From science and now the Internet, we know that things stand together in complex interrelated relationships. This 'truth' brought science and society to examine the nature and form of those relationships. This supports the concept of a God that is everywhere and not silent. Without this certainty the pursuit of knowledge is at best meaningless. This panentheistic theology can guide religions in establishing societal ethics and law.

If global consciousness mimics human consciousness what would it look like? How would it relate an individual to society and to God? Earlier a hierarchical societal view was presented as analogous to human consciousness.

Consider a hierarchy based on human consciousness augmented by the Internet. If we are made in God's image, by taking Beer's Cybernetic view of organization based on the human nervous system, we can create a systemic hierarchy of | how human consciousness relates to the mind of a panentheistic God.

God's Domain
Intelligence↔Cohesion
Our Universe & Others
Intelligence↔Cohesion
Planet Earth Noosphere
Intelligence↔Cohesion
Society
Intelligence↔Cohesion
Community
Intelligence↔Cohesion
Individual
Intelligence↔Cohesion

6 Levels of Panentheistic Consciousness

Using Beer's cybernetic model, consider the next figure where Systems 3, 4, 5 represent God's consciousness:

- System 5: Influence toward ultimate (omniscient, omnipresent, omni-beneficent) harmony.
- System 4: Scan, plan, and adapt to universal, galactic, planetary changes and conscious creature problems. God's "external" environment is also "internal." [26]
- System 3: Internally regulate and optimize physical laws; harmonize conflicts in conscious creatures. God is aware of every molecule in His universes.
- System 2: Coordinate events and resolve conflicts.

---

[26] To imagine how God's domain (Isvara) can be inside/outside the universe, consider a mathematical concept from topology: a 'Klein" bottle (ref: kleinbottle.com) with no inside or no outside.

- System 1: Maintain the universe (physics) and its biological creatures (genetics).

Information flowing in this model is:

- IPP_1 – identify novel ideas and solutions in problematic environments of God's conscious creatures.
- IPP_2 – coordinate and enable universal activities in support of physical, biological and social laws and individual and societal values.
- IPP_3 - support adaptation by communicating between the Cohesion (3) and Intelligence (4) functions using ideas and eternal objects in God's primordial nature.
- IPP_4 - summarize and monitor IPP_3 for effective value generation and harmonizing activities in God's consequential nature.

To be clear, this only speculates on the mind of God. It reflects past solutions to societal economic and political problems using systems theory. It frames a communications system for religions and nations as to how they communicate to individuals, communities and each other.

Multiple universes in cosmology and eleven dimensions in physics are modern day mythology. Randomness in evolution and quantum mechanics is not causation. No one can explain infinity in mathematics. An imagined panentheistic God is necessary in a philosophy of science.

Value

Value from religion is guided by an ecumenical theology.	
• 21st Century Theology • Theology Paradigm • Purpose • Value/Ethics	• How should religions body of knowledge be extended? • By what ethics will people live their lives? • What is our purpose: vocational, familial, societal, spiritual? • What will a science of religion look like and do? • How can orthodox churches make change?
Aided by Whitehead and others, this book attempts to present a limited view of God's truth and renew the theological quest for truth. It contends that God interacts with humankind through natural laws (e.g. Biology, Physics, Sociology). The book proposes that theology must evolve to provide a system and discipline that provides moral guidance to our present humanistic secular western society or risk falling back to 16th century theocracy. To prove God exists, philosophy and science have proved inadequate (e.g. refutations of Augustine, Aquinas, Descartes, Kant 'proofs of God's existence'). What emerged is the synthesis proposed earlier as a narrative explaining how and why religion: 1] Benefits Individual well-being, 2] Precipitates and validates revelation, 3] Was used to charge societies with value, 4] Provided the foundation for science and belief in rational thinking, 5] Can evolve based on an ecumenical theology that guides all religions and derives ethics and values for individuals, communities and societies.	

As atheism and secular humanism imbed their theocracy on society, they systematically exclude religions from the public square; this is not religious freedom. Until established religions accept an ecumenical theology to influence societal ethics, atheism's corrosive effects on society, community and individuals will continue.

Ethics for society and individuals is not relative. An Ecumenical theology can guide religions and establish societal ethics and law. Religion must provide a moral compass for the 21st century.

Thomas Jefferson (not a Christian) wrote that Jesus Christ's words represent the Word of God: *"sublime ideas of the Supreme Being, aphorisms and precepts of the purest morality and benevolence, sanctioned by a life of humility, innocence, and simplicity of manners, neglect of riches, absence of worldly ambition and honors, with eloquence and persuasiveness."*

Morality under the law was summed up by Jesus with what some now refer to as the Golden Rule: "All things whatsoever you would that men should do to you, do you even so to them: for this is the law and the prophets" (Matthew 7:12). It is expanded from an individual rule to a guide for law by Immanuel Kant in his Metaphysics of Morals: "Act only on that maxim whereby you can at the same time will that it should become a universal law."

For Christians, the teachings of Jesus Christ guide the duties of moral and ethical behavior. Individuals must redefine those duties for themselves. Unfortunately, many Christian communities devalue the sublime ideas of other religions and science and their application to our individual and national conscience.

Science and Christianity's conflict is partially explained by a literal interpretation of the Bible, which became prevalent in the mid-1800s after the Protestant Reformation's emphasis on the Bible as the only authoritative source concerning the ultimate reality. Catholicism's St. Augustine explicitly opposed a literal interpretation of the Bible whenever the Bible conflicted with Science.

To base a modern theology on Christianity would be to ignore the majority of the world's faiths. A universal theological community is suggested based on Kuhn's scientific community whose members would include scholars from philosophy, science, and the world's major religions as well as atheists and agnostics. Such a community would adopt a formal argumentation for moral consciousness to vet and validate old revelation in ancient scriptures and new revelation.

In Religion in the Making, Whitehead wrote, "The great rational religions are the outcome of the emergence of a religious consciousness which is universal, as distinguished from tribal or even social. Because it is universal, it introduces the note of solitariness. Religion is what the individual does with his solitariness."

Swami Vivikananda's first concept of universal religion is the one Eternal Religion that represents the religious consciousness of humanity, which manifests itself in different places as different religions. The religions of the world are various phases of one Eternal Religion.

Vivikananda's second concept of universal religion is the co-existence of all the religions to form a whole; its basic principles are:

- Recognize and respect the unique features of each religion and its right to retain its individuality.
- World Religions are not contradictory, but complementary. '...religions are different forces in the economy of God, working for mankind.'
- World Religions could interact with one another for humanity's welfare; in a perpetual interreligious dialogue, or mutual sharing in a spirit of acceptance.

Vivikananda's third concept of universal religion is an integral view of Life and Reality. It is man's personal quest to transcend his limitations, and find ultimate meaning in life in his or her own religion. However, it also has a collective aspect as a fivefold harmony:

- Harmony between the sacred and the secular by sacralizing the secular, by divinizing the whole life.
- Harmony between science and religion as a single quest of man to know the ultimate Truth.
- Harmony between love for man and God by seeing God in man. Man's true nature is inseparable from God; love expresses the spiritual oneness of all humanity in God.
- Harmony between contemplative and active life by the practice of meditative self-awareness where distinctions between contemplative and active life disappear.

- Harmony in personality development of every person's four faculties: thinking, feeling, willing and work.

For Vivekananda religion transforms human life into Divine Life. It converts every thought, feeling, and action into a spiritual discipline and sanctifies one's whole life.

The Pontifical Academy of Sciences might represent a model of a collaborative universal theological community. It was founded in 1603 and later re-established in 1936 by Pope Pius XI. In 1979, John Paul II emphasized the role and its goals:"... today eminent scientists are members, without any form of ethnic or religious discrimination, is a visible sign, raised amongst the peoples of the world, of the profound harmony that can exist between the truths of science and the truths of faith..."

According to St. Augustine, God reveals himself to humans by divine accommodation (man's current ability to understand). The revelation of the past 1500 years including the arts and sciences represent man's more comprehensive view of God's natural laws. A framework is required that represents disparate religious dogmas as well as scientific insight and human and societal ethics. It is summarized below:

Truth	Dogma	Word	Worship	Program
• "Sacred" texts • References • Candidate References	• Theology / Philosophy • Other Reference	• Religious text • Liturgy	• Service • Sermons • Fellowship • Music	• Behavior • Ministry
Search for Truth	**Extend Dogma**	**Extend the Word**	**Expand Community**	**Operate Programs**
• Kuhn derived theological community • Religious scholars	• Religious Scholars • Ministers	• Educators • Ministers	• Ministers • Stewards • Leaders	• Ministers • Believers • Missionaries

While the last four processes would be unique to a particular religion, the first process, *Search for Truth*, considers all major religions as well as new revelations (dynamic patterns) and Pirsig's four static systems of patterns: inorganic, biological, social, and intellectual. As caretakers of the planet, people are responsible for preservation of habitat and species as well as creating ethical standards and laws for social and individual behavior.

Consider the last four processes in terms of the Christian faith, community and worship:

Table	Interpret the four processes in terms of Christianity
Extend Dogma	Theology/Philosophy: • St. Augustine: Contrasts the perfect City of God with man's world. • St. Thomas Aquinas: Attempted to resolve Greek philosophy and science with Christian theology. • Martin Luther: Wanted the Catholic Church to walk their talk; he said an individual could come to God through knowledge of the Bible and not through the church hierarchy. • John Calvin: Said that one is saved by the grace of Jesus, and as a consequence of life in God, one is required to live a good life and develop a productive profession. • Whitehead and Hartshorne laid the foundation for process theology. • Scientific method has been used to show the fallibility of the Biblical view of cosmology and creation. • Literature has been used to show the benefits of Christian values and crimes of religion. • Art and literature has been used to illustrate God's relationship to us.
Extend the Word	• Religious text: Bible - Man's attempt to relate God's truth in the history of the faithful. • Parables of Jesus: Provide a guide for Christian behavior.
Expand Community	• Liturgy: The script for worship. • Music: Reinforces church dogma and makes worship aesthetic. • Fellowship: Reflects the grace and love of the Christian community. • Service Structure: Provides a community ritual with music, prayer, liturgy and sermon.
Operate programs	• Behavior: Christians are expected to be honest, forgiving and loving to all people not just their immediate community. • Ministries: By example and actions, Christians are expected to bring others into their community.

In his book, *Moral Consciousness and Communicative Action*, Jürgen Habermas contrasts moral philosophy with a developmental psychology of moral consciousness from Lawrence Kohlberg's *Essays on Moral Development* (San Francisco, 1981) in six stages:

1] Egocentric perspective: ignores other viewpoints; judges action based on physical consequences rather than psychological differences of others; confuses authority's perspective with one's own.

2] Individualistic perspective: separates one's own interests and viewpoint from those of authority and others; integrates conflicting individual interests by exchange of services, a need for the other person's goodwill.

3] Relationship to other individual's perspective: is aware of shared feelings, agreements, and expectations which take primacy over individual interests; relates viewpoints through the "Golden Rule," but does not consider a general "system" perspective: e.g. Jesus' Good Samaritan.

4] Societal viewpoint perspective: takes the viewpoint of the system that defines roles and rules; considers individual relations as a place in the system (Kant).

5] Prior to society perspective: is aware of values and rights prior to social attachments and contracts; recognizes the conflict of moral and legal viewpoints and finds it difficult to integrate them.

6] Moral viewpoint that derives social arrangements perspective: recognizes the basic moral premise of respect for other persons as an end, not a means.

Habermas takes us only to 4] by surmising that theory depends on communicative action among individuals (theological community). Argumentation starts in the particular "lifeworld" - conventional, habitual, concrete, self-evident, certain - and moves to a hypothetical attitude that embraces reflection, self-regard, and abstraction.

One might ask by what authority does 3] through 6] come from. 3) is stated originally in the old testament, but is attributed to Jesus in Matthew 7:12 – "Therefore all things whatsoever ye would that men should do to you, do ye even so to them: for this is the law and the prophets." 5] and 6] in Eastern philosophy is resolved by intuition of Reality in transcendent consciousness with Isvara.

Western theology is not confused either, authority comes from God; Western philosophy gets confused with its emphasis on the scientific method, absent value, ethics and aesthetics, and thinks argumentation is the answer and not validation in a theological community that acknowledges transcendent intuition in oneness with God.

New transcendent perceptions must be vetted through the proposed theological community. This is the world of general principles, about which people argue (reason) and win or lose on the basis of the "force" of argument rather than force of influence, superior might or invoking God. Habermas sums up: "justice can be gleaned only from the idealized form of reciprocity that underlies discourse."

Habermas' summary missed a critical point. As science progresses through divine and human revelation, it can significantly change society and societal norms (e.g. planes, nuclear bombs, gene splicing, etc.). Kant says, one has *a priori* knowledge of duty and God. Whitehead's primordial nature of God requires the discovery of more eternal objects. Adaptation by a society may require divine or human revelation to extend the current view of justice, which is validated "from the idealized form of reciprocity that underlies discourse." Perhaps then, a society governed by the laws of level four Moral Development can be sustained.

World Economic Forum's Global Competitive Index seeks to measure the prosperity for citizens in 138 countries. As we evolve to a world economy, we are evolving to a global consciousness without a global conscience. The challenge for the world's major religions is to collaboratively be its conscience and to demonstrate the values and ethics for the emerging global society.

This ecumenical group must develop new measures for the new world socio-cultural system that go beyond the WEF's measures for social and environmental wellbeing. They must measure elements of moral wellbeing that reflect a level four Moral Development. This involves force of argument rather than force of influence, or superior might. It may invoke God's consequent nature.

In his book, *The Varieties of Religious Experience*, William James reconciled scientific with religious knowledge by proposing that if religious beliefs could be validated as useful in helping people cope with difficult emotions or moral decisions, it would indicate a certain truth for those beliefs. James sums up religious life as belief: that 1] the visible world is part of a more spiritual universe from which it draws its chief significance; that 2] union or harmonious relation with that higher universe is our true end; 3] that prayer or inner communion with the spirit — "God" or "law"--is a process wherein work is really done, and spiritual energy flows in and produces effects, psychological or material.

James concludes that the following are psychological characteristics of religious belief: 4] includes a new zest for life, and takes the form of lyrical enchantment or appeal to earnestness and heroism; and 5] provides an assurance of safety and a temper of peace, and a preponderance of loving affections in relation to others.

Krishananda criticizes James' pragmatism as not reality but merely a network of evidences of the senses. As a psychologist, James trusts psychological functions that he identifies with consciousness. Mental consciousness is a stream, a flow, a becoming. Knowledge of a stream cannot itself be a stream. Utility is no test of truth. What constantly changes is not ultimate truth; change moves toward 'something.' Ultimate truth is not a means to an end, but an end in itself; it is partially revealed in consciousness. James maintains that we create reality every moment as a stream of perceptions and ideas with relations between them. No relation of ideas within is possible without an indivisible Self.

Krishnananda misses the point that James harmonious expression of wholeness agrees with the Vedanta's notion of keeping the law of Isvara so that the world does not recoil upon you. James claims that one's belief can be evaluated to some extent in that the divine will be evident in the behavior, demeanor and mental health of the believer if his belief keeps Isvara's law.

An ecumenical theology does not need to waste time trying to prove that God exists or even worse trying to prove God does not exist. Denying God's existence in history, sociology, psychology, and anthropology denies the obvious and has resulted in weak paradigms that have not improved the human condition.

Any analysis of humanity, community, and society past and present must acknowledge that belief in God is fundamental to evolution of human consciousness. Over 80% of the people on the planet belong to one of the major religions. Ninety-five percent have some kind of personal relationship with God that they call spiritual.

Religion and belief in God has been a key factor in the reduction of crime and addiction and in the rehabilitation of criminals and addicts. If we are to live in a free peaceful society, we must fully exploit and support people's need for transcendence, to be part of a 'communion of saints,' and to have a relationship with God.

The social sciences must not deny God, because society needs viable and ethical direction that benefits rather than erodes our humanity by denying the expression of God in the public square. They can rename God to get past whatever negativity that word suggests for them (e.g. Crusades); call God: He/She/It, First Cause, Creator, The Force, or even small 'g' god. *Like belief in infinity in mathematics, acknowledging belief in god is fundamental to the human condition and is critical to analyzing it.*

A panentheistic systems view of god has been presented. This view permits speculation about god's processes and information requirements based on humanity being created in the "image of God." This view is not God. This is the author's concept to present a framework that minimally contradicts known facts, and the elements of science and major world religions.

This god of an ecumenical theology will require certain qualities as defined by Whitehead if the analysis of the human condition is to become a science. The major religions of the world will assign qualities beyond God's primordial and consequent nature, which is OK provided a 'systems" view of God lies at the core of societal and human analysis and religious belief.

This "religion" started with the synthesis of ideas from Kuhn, Pirsig, Northrop, Habermas, Whitehead, James, Vivekananda, Krishnananda and possibly Beer to build an infrastructure that promotes seeking for the truth and assisting each of its followers to build a personal faith.

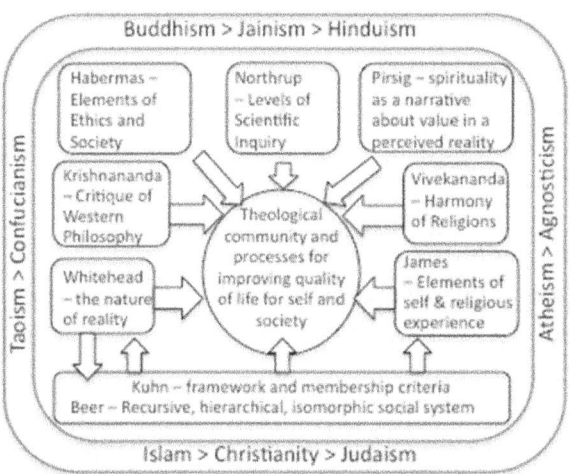

No God and no religion in the public square have worked to the detriment of society, but religions cloaked in ancient dogmas haven't helped. The figure synthesizes this book's proposed ideas to lead us out of this quagmire.

The elements of a social science must acknowledge and develop analytical structures for a person to love self, family, neighbors, community, country, world and God. Answers to modern day problems must be sought with the belief that God's eternal objects await discovery.

A minimal theology upon which a community of scholars begins argumentation might include:

- One God (He/She/It) who embodies all knowledge and maintains our universe and us (or universal law).
- Humans' theological purpose is to understand, qualify and quantify God's natural laws and eternal objects.
- A human's life purpose is to love and ethically care for family, community, country and world.
- The community's purpose is to define a theology for humans to realize their purpose and to define values that support level 4] Societal Viewpoint Perspective.

The theological community must:

- Add to scripture and worship from a universal theology.
- Teach communities how to understand scripture.
- Counsel individuals on how to achieve transcendence.
- Counsel people to develop healthy habits of "virtue."
- Counsel societal leaders in ethics and moral value.
- Counsel individuals in how to love and serve oneself, family, others, community, country, world and God.

A theological community must help individuals answer a variety of questions; here are a few with sample answers:

- Why are you here? I am here to discover how I might serve in harmony with God's intended purpose for the community, the world, the universe, and myself.
- What do you value? I value that which benefits me in the context of family, community, society and the world. I value my God-given consciousness that allows me to seek knowledge of myself, the world, and God.

- What is your religion, what do you believe and who is God? I believe in a God that harmonizes the universe; my beliefs are aware of many religious dogmas, but I will not simply adopt one absent reason.
- What is your code of ethics? I accept a standard of behavior that best serves myself in community. I adhere to this while attempting to act in a way that does the most good for community, environment and myself.
- How will you relate to yourself as an individual, to God and the Universe? I will treat all three with the love and awe that each deserves.
- How will you relate to others in community and society? I will treat each with dignity and respect.
- How are you going to contribute? I will discover what gifts I have been given and attempt to refine those capabilities while applying them for my personal, family, community and environmental welfare.
- How will you become a good parent, citizen and friend? I will be someone people trust. I do what I say I will do.
- Why and what must you learn about community, society and history? I must learn enough to be an informed citizen in a democracy to protect its freedoms.
- How will you protect people, earth, country and way of life? I will learn the relevant facts about each topic – religion, science, politics, economics, and culture; decide what is critical for each one; and work toward and/or support others who work toward those ends.
- How should government serve us and how is it paid for? Government should protect its citizens from within and without and require its citizens to meet minimum standards of citizenship to vote in a democracy. Recognizing that government burdens businesses and citizens with taxes, it should minimize its functions.
- How is society protected? Society is protected by its military, its police force, its robust economy and, in a democracy, the intelligent participation of citizens.

- What is law and how should it be administered? Law protects responsible citizens in a society from evil behavior as enforced by police, courts, punishment and prison. It should not be a way to force 'proper' behavior lest a democracy become a fascist state.
- What is ethical: abortion, stem cell research, cloning, etc.? Ethics should not be based on convenience or political activism. It must be based on truth, God's eternal objects and natural law, and evolve from an ecumenical theological community.
- How much more don't you know that you don't know you don't know? Seven billion people may not put a dent in the infinity of unknown eternal objects, but humans have illuminated millions of them. It is our job as humans to never be intimidated by the breadth and depth of our ignorance in searching for God's truth.

Religion must not only guide individuals to lead virtuous lives, but also must provide society and the global community with a moral viewpoint "that recognizes the basic moral premise of respect for other persons as an end, not a means." Thought leaders of the world religions must acknowledge the value that other religions afford and together identify a universal system of values to guide individual and societal behavior.

This book proposed an ecumenical theological system, process and community. This system must help members of all religions in their spiritual relationship with God and each other. It must show how faith leads to rational ways that each of us can work together to use our God given consciousness to progress toward our ultimate purpose as individuals of society and the global community.

Some of us would like to know why we have been given the gift of consciousness and a need to know God and what God thinks (the rest are details).

Our reality is driven by flickerings of charged quanta in a cosmic curved space-time, which we comprehend only with mathematical abstractions, imagination and personal transcendence.

"Poised midway between the unvisualizable cosmic vastness of curved space-time and the dubious, shadowy flickerings of charged quanta, we human beings, more like rainbows and mirages than like raindrops or boulders, are unpredictable self-writing poems: vague, metaphorical, ambiguous, and sometimes exceedingly beautiful."
- Douglas Hofstadter

Religion must help us to live a life of multiple purposes. Religious transcendence allows an individual to "act with honor in virtuous serenity." Consciousness and transcendence are too great a gift to squander. None of us are perfect, but by God's grace that's OK. Theology like science requires community and reasonable belief.

The order of the world is no accident. There is nothing actual which could be actual without some measure of order. Religious insight is the grasp of this truth: That the order of the world, the depth of reality of the world, the value of the world in its whole and in its parts, the beauty of the world, the zest of life, the peace of life, and the mastery of evil, are all bound together - not accidentally, but by reason of this truth: that the universe exhibits a creativity with infinite freedom, and a realm of forms with infinite possibilities; but that this creativity and these forms are together impotent to achieve actuality apart from the completed ideal harmony, which is God.
- Alfred North Whitehead

Bibliography

- Aquinas, Thomas, *Short Summa Theologica*, 1261, Sophia Institute, 1993.
- Asimov, Isaac, *Asimov's Guide to the Bible*, Wing Books, 1969.
- Augustine, *City of God*, 410, Doubleday, 1958.
- Beer, Stafford, Brain of the Firm, John Wiley and Sons, 1981.
- Berkeley, George, The Principles of Human Knowledge, 1710.
- *Bible*, Old Testament Redactions 950, 850, 750, 400 BCE, New Testament Assembly, 200.
- Brennan, Sheilah O'Flynn, *Perception and Causality: Whitehead and Aristotle*, Process Studies, pp. 273-284, Vol. 3, Number 4, Winter, 1973 .
- Cobb, John B. Jr. and Griffin, David Ray, *Process Theology*, Westminster Press, 1976.
- Collins, Francis S., *The Language of God*, Simon and Schuster, 2007.
- Confucius, *Analects of Confucius*, 475 BCE.
- Cooper, John W., 2006, Panentheism The Other God of the Philosophers: From Plato to the Present, Grand Rapids, MI: Baker Academic.
- Dawkins, Richard, *The God Delusion*, Houghton Mifflin Company, 2006.
- Descartes, Rene, *Rules for the Direction of the Mind*, 1629, translated by E.S. Haldane and G.R.T. Ross, Cambridge University Press, 1937.
- Deutsch, Eliot, *Advaita Vedanta: A Philosophical Reconstruction*, University of Hawaii Press (Dec 1980).
- Dewey, John, *How We Think*, D.C. Heath & Co., 1910.
- Earhart, H. Byron, *Religious Traditions of the World*, Harper Collins, 1993.

- Gödel, Kurt, On Formally Undecidable Propositions of Principa Mathematica and Related Systems, 1931.
- Greene, Brian, *The Elegant Universe*, Vintage Books, 2003.
- Habermas, Juergen, Moral Consciousness and Communicative Action, MIT Press, 1990.
- Harris, Sam, *End of Faith*, W. W. Norton Co., 2004.
- Hartshorne, Charles, *The Divine Relativity*, Yale University Press, 1948.
- Hitchens, Christopher, *God is not Great*, Hachette Book Group, 2007.
- Hofstadter, Douglas, *I Am a Strange Loop*, Perseus Book Group, 2007.
- Hume, David, Dialogues Concerning Natural Religion, Clay Ltd, 1779.
- James, William, *Talks to Teachers on Psychology*, Harvard, 1892.
- James, William, *The Varieties of Religious Experience*, Gifford Lectures, 1902.
- Jefferson, Thomas, *The Jefferson Bible*, Beacon Press, 1989.
- Kant, Immanuel, *Metaphysics of Morals*, Categorical Imperative, 1780.
- Krishnananda, Saraswati, *The Mandukya Upanishad*, The Divine Life Society.
- Krishnananda, Saraswati, *The Philosophy of Life*, The Divine Life Society.
- Kuhn, Thomas S., *The Structure of Scientific Revolutions*, The University of Chicago Press, 1962.
- Lewis, C. S., *Mere Christianity*, Macmillan Publishing Company, 1943.
- Locke, John, Concerning Human Understanding, Holt, 1690.
- Machiavelli, Niccolo, *The Prince*, Oxford University Press, 1935.

- Pagels, Elaine, *Beyond Belief*, Harper Collins Publishers, 2003.
- Pirsig, Robert, *Lila - An Inquiry into Morals*, Bantam, 1991.
- Plato, *Dialogues of Plato, Meno and Theaetetus*, 387 B.C., translated by B. Jowett, London, Sphere, 1970.
- Polkinghorne, John, *The Faith of a Physicist*, First Fortress Press, 1996.
- Russell, Bertrand, *Why I am not a Christian*, Simon and Schuster, 1957.
- Schmidt, Wilhelm, *The High Gods in North America*, Oxford: Clarendon Press, 1933.
- Sherburne, Donald W., *A Key to Whitehead's Process and Reality*, University of Chicago Press, 1929.
- Sneddon, Andrew, Master Thesis: *A Process Analysis of Quality*, Ottawa University, 1995.
- Stark, Rodney, *Discovering God*, Harper Collins, 2007.
- Stenger, Victor J., *God: The Failed Hypothesis*, Prometheus Books, 2007.
- Strobel, Lee, *The Case for Christ*, Zondervan, 1998.
- Taylor, Charles, *A Secular Age*, Harvard University Press, 2007.
- Tsu, Lao, *The Way of Life*, 600 B.C., translated by Witter Bynner, Lyrebird, 1972.
- Tsu, Lao, *Tao Te Ching*, 600 B.C., translated by Gia-Fu Feng and Jane English, Random House 1972.
- Whitehead, Alfred North, *Religion in the Making*, 1926.
- Whitehead, Alfred North, *Process and Reality*, 1929.
- Wolfram, Stephen, *A New Kind of Science*, Wolfram Media, Inc. 2002.
- Wright, Jonathan, *God's Soldiers*, Doubleday, 2004.

www.ingramcontent.com/pod-product-compliance
Lightning Source LLC
LaVergne TN
LVHW011424080426
835512LV00005B/254